Molecular and Cellular Biology

Cohesive, Concise, yet Comprehensive Introduction for Students and Professionals

Paul Sanghera, Ph.D.

Molecular and Cellular Biology: Cohesive, Concise, yet Comprehensive Introduction for Students and Professionals

Published by
Infonential, Inc.

Technical Editor: Dr. John Serri
Copy Editor: Arlene Guanlao

Copyright © 2015 by Infonential. All rights reserved. No part of this book may be reproduced, stored in a retrieval system, or transmitted in any form by any means, including electronic, print, photocopying, recording, scanning, or otherwise, without written permission from the publisher.

Limit of Liability/Disclaimer of Warranty: While the publisher and author have used their best efforts in preparing this book, they make no representations or warranties with respect to the accuracy or completeness of the contents of this book and specifically disclaim any implied warranties of merchantability or fitness for a particular purpose. No warranty may be created or extended by sales representatives or written sales materials. The advice and strategies contained herein may not be suitable for your situation. You should consult with a professional where appropriate. Neither the publisher nor the author shall be liable for any loss of profit or any other commercial damages, including but not limited to special, incidental, consequential, or other damages.

ISBN-13: 978-1484016091
ISBN-10: 1484016092

To

Biology students and professionals all across the world

About the Book

Molecular and Cellular Biology: Cohesive, Concise, Yet Comprehensive Introduction for Students and Professionals

In this book, Dr. Paul Sanghera, the best-selling author of several books in science and technology, provides cohesive, concise, yet comprehensive coverage of the key concepts of molecular and cellular biology in an accessible way. The book presents material in a logical learning sequence: each section builds upon previous sections and each chapter upon previous chapters. All concepts – simple and complex – are well-defined and clearly explained the first time they appear. There is no hopping from topic to topic and no technical jargon without explanation. This book is useful for both students and professionals in biology. Students can use the distilled information in this book to excel in their assignments and exams. Even though this book is self-contained, it works as a great supplement to any textbook in general biology. Professionals in a biology-related field can use it for a quick reference or for a quick review of fundamental concepts, whereas newcomers can use it as their gateway into the field to quickly ramp up to speed.

The chapters in the book have the following special features:

- **Note.** A *Note* is used to present additional helpful material related to the topic being described or to emphasize a concept.

- **Caution.** A *Caution* is used to highlight a point which either is crucial or may not fit into the commonsense framework of things.

- **Think About It.** This feature presents questions or simple problems with answers and solutions to emphasize critical concepts.
- **Problems.** Problems are presented with solutions to explain mathematical concepts.
- **Fascinating Fact.** This feature singles out biological facts which are simply amazing.
- **Review Questions.** Review questions with answers are presented at the end of each chapter in order to enable you to test your knowledge and detect your strengths and weaknesses.
- **Glossary.** This feature presents quick access to key terms.

Enjoy!

About the Author

Dr. Paul Sanghera, an educator, scientist, engineer, technologist, author, and entrepreneur, has a diverse background and experience in multiple fields, including physics, chemistry, mathematics, computer science, and biosciences. He holds a Ph.D. in Physics from Carleton University, Canada; a Master of Engineering degree in Computer Science from Cornell University, U.S.A.; and a B.Sc. from India with a triple major in physics, chemistry, and math. He has comprehensive, cross-disciplinary, cross-continental, and diverse experience in research, teaching, and learning. He has taught a wide spectrum of science and technology courses all across the world, including San Jose State University, U.S.A.; Carleton University and Simon Fraser University, Canada; and Indian Institute of Technology (IIT), India. He has authored and coauthored about 150 science research papers on subatomic particles of matter that have been published in well-reputed European, American, and international research journals. At world-class laboratories, such as CERN in Europe and the Nuclear Lab at Cornell, he has participated in designing and conducting experiments to test quantum theories and models of subatomic particles and thereby has contributed to testing the Standard Model of the universe.

Dr. Sanghera was a lead engineer at the first Internet company, Netscape Communications Inc. He has been at the ground floor of several start-ups in the Silicon Valley in California and elsewhere, and he has contributed to the development of state-of-the-art technologies such as Novell's NDS, the first computer network management system, and one of the first commercial web storefronts at Weborder Inc.

Dr. Sanghera is the author of about two dozen books in science, technology, and project management. His current research interests involve topics in biomolecular engineering and bioinformatics.

Contents

About the Book — v

About the Author — vii

1. Defining Life — 1
2. Chemistry of Life — 19
3. Molecules of Life — 47
4. Anatomy and Physiology of Cells — 69
5. Cell Membrane: The Supermolecule of Life — 93
6. Cellular Energy — 117
7. The Cell Cycle — 155
8. Classical Genetics — 177
9. Molecular Genetics — 197
10. Biotechnology — 239

Glossary — 273

Credits and Acknowledgments — 283

Chapter 1

Defining Life

1.1 Life and Biology: The Big Picture

Science is the study of nature. Nature comprises of everything in the universe except the things made by us, the artificial things. Scientists discover the laws of nature, also called the physical laws, which govern nature. We often use the knowledge obtained from these discoveries to make artificial things. Nature consists of living and nonliving entities. Biology is the branch of science that studies living entities and their related processes. The science of biology draws on other scientific disciplines such as physics (biophysics), chemistry (biochemistry), computer science (bioinformatics), and engineering (bioengineering).

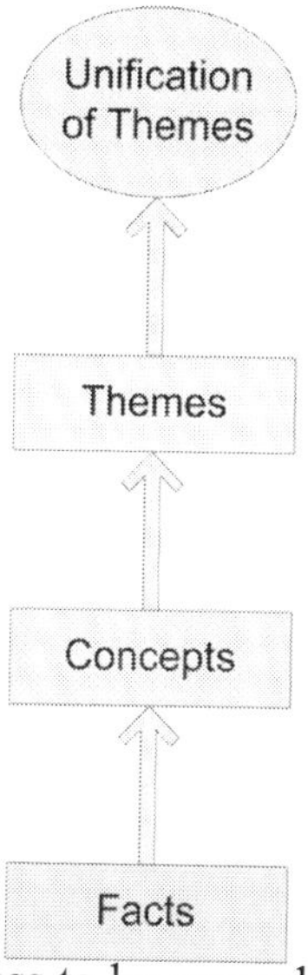

Science is not just an accumulation of facts, but it is also a process through which scientists accumulate knowledge. This process of scientific discovery includes hypothesizing and experimentation.

Figure 1.1 Relationship between facts, concepts, and themes.

Although it is important for your professional success to know and remember the technical terminology of a given discipline, it is even more important to develop a comprehensive

Molecular and Cellular Biology, by Paul Sanghera
Copyright © 2015 Infonential.

Molecular and Cellular Biology

understanding of the bigger and ordered framework of knowledge and to understand the concepts behind the facts and the themes that run through multiple concepts, as illustrated in Figure 1.1.

Theme. A theme is a central or unifying idea. It runs across and explains or unifies multiple concepts. For example, a theme is that life evolves. It explains the two apparently opposing concepts of unity of life and diversity of life.

Concept. A concept is an idea that explains multiple facts. For example, the enormous number of facts that indicate all organisms do similar things is explained by the concept of underlying unity in the fundamental structure and functioning of life. The enormous number of facts that indicate there are significant differences among different species is explained by the concept that life is diverse. Remember that there is some overlap between a concept and a theme. A unifying concept (or idea) is a theme.

> Note. An organism is any living individual with one or more cells such as a bacterium, animal, or plant.

Facts. These are scientific observations. For example, consider the fact that different organisms do similar things: all of them respond to their environment, reproduce, grow and develop, and so on. Consider another fact: different species of organisms are different from each other in many morphological and physiological aspects.

Multiple themes can also be unified into a central theme or a core theme. For example, the theme of evolution can also be looked upon as a unified core theme because it runs through many other themes in biology. It explains the unity behind the diversity of life.

In this chapter, we will explore the characteristics of life, different organizational levels of life, and major themes of biology. We will begin this exploration by answering an important question: what is life?

1.2 What Is Life?

At first, this may sound like a silly question because in our daily lives, we generally have a very good idea of what life is and what it is not. For example, even a child knows that a lush green tree is alive and a rock is not. To be more specific, life is defined by a set of characteristics that living things have, which include the following:

Metabolism and Growth. Organisms metabolize and grow. *Metabolism* is a set of chemical reactions that includes *anabolic* (consuming energy to synthesize a complex molecule) and *catabolic* (releasing energy by breaking down a complex molecule) pathways. Metabolism provides the energy required to accomplish growth. In order for an organism to grow, it takes in suitable material such as food from the environment, breaks it down if necessary, and then reorganizes it into a needed structure. For instance, when you eat a meal and digest it, your body transforms it into more of yourself (energy and molecules required to maintain life) by using metabolism.

Nonliving entities do not metabolize and grow.

Response to the Environment. An organism interacts with its environment, senses it, and responds to changes occurring in the environment internal or external to its body. Energy is generally used in this response. For example, you feel hungry (sensing the internal environment), and you eat (response). Different responses constitute different behaviors of different organisms. For example, in response to the disturbance caused by you, an organism may run away from you, while another organism may attack you. These two responses represent two different behaviors.

The responsive behavior of a living organism is generally active as opposed to the passive (nonresponsive) behavior of

Molecular and Cellular Biology

a nonliving entity. For example, when you slip, you actively try to avoid falling or to fall in a safer way, which is different from the passive behavior of a book that drops from your hands.

Reproduction. Organisms reproduce their own kind; that is, they reproduce their own species. Reproduction is of two types: *asexual reproduction* and *sexual reproduction*. In asexual reproduction, offspring arises from only one parent and therefore inherits the genes of one parent only. It is like producing a clone. In sexual reproduction, however, offspring acquires half of its genes from each parent. Therefore, sexual reproduction is more complex and gives rise to more diversity and new unique properties and species over time. A molecule called deoxyribose nucleic acid (DNA), the carrier of genes, is the basis of growth and reproduction. We will explore this topic in more detail further on in the book.

Only organisms have DNA; nonliving entities do not.

Homeostasis and Regulation. Another defining feature of life is *homeostasis*, which is a steady physiological state in which certain parameters are kept within a tolerable range in the body of a multicelled organism. For example, homeostasis keeps internal temperature and pH value (a measure of the concentration of hydrogen ions or H^+, which are practically protons) within a very narrow tolerable range. This steady state is accomplished by sensing and responding to changes in the environment, which includes regulating some processes. For example, *thermoregulation* in an organism helps maintain a constant internal body temperature even if the external temperature changes. Many types of regulation are performed through a *feedback mechanism*, which is a mechanism that involves signaling molecules called hormones.

Interaction and Energy Exchange with the Environment. All living organisms interact with their environment in many ways. For example, they take energy and nutrients from the environment, which are necessary for their growth and survival. Some organisms prepare their own food from this energy and are called autotrophs. For example, plants prepare their own food (glucose) from carbon dioxide, sunlight, and water by using a process called photosynthesis. Other type of organisms, called heterotrophs, ingest food and obtain energy from it through metabolism. Animals and humans are heterotrophs. The environment can change a population of living organisms as they adapt to it, and organisms can also alter their environment. However, nonliving things cannot alter the environment on their own. The study of the interaction between the environment and life at different organizational levels is called *ecology*.

Evolution. Species of organisms evolve as populations adapt to their environments. Evolution is the process in which species evolved by descending with modification from a common ancestor. Through evolution, a population may evolve into a new species while responding and adapting to a changing environment.

Unity and diversity are two apparently contradictory features of life. An example of unity is that all organisms are made of cells. An example of diversity is that the total number of different species of organisms are estimated in millions; even two organisms of the same species such as two humans don't look exactly alike (except perhaps the identical twins). Evolution is the central theme of life that explains the unity behind diversity. When there is a change in the environment, the population changes to one that can better adapt, respond, metabolize, grow, and reproduce in the new environment. The changed population has developed characteristics (or traits) that the previous population did not have. This is how new species evolve. This process allows evolution to give rise to greater diversity over time.

New emergent properties at each organizational level. All living organisms are composed of one or more cells, the basic units of life. Starting from cells, organisms have

Molecular and Cellular Biology

different levels of organization, illustrated in Figure 1.2. At each level of organizational complexity, life is more than the sum of its parts. To be specific, at each level of organization, some new properties emerge that are not part of any of the components of the system at that level. These properties emerge from the interactions between parts of the system at that level (such as molecules in a cell or cells in an organ) and the structures that result from these interactions. For example, a cell is composed of molecules, but life appears only at the cellular level and not at the level of any of the molecules of the cell. The cell performs functions that none of its molecules can perform just by itself.

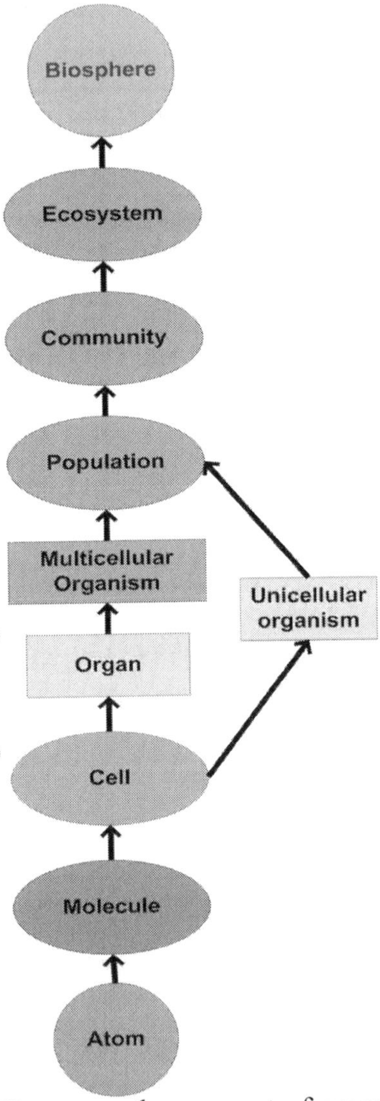

Different organizational levels with emergent properties are very important to understand life. However, the concept of emergent properties is not unique to life, as it also applies to nonliving systems. For example, your car provides you transportation, a property which none of its parts alone can offer.

Figure 1.2 Organizational levels of life

6

So, new properties emerge at each organi·
life. However, what exactly are these organiza
us explore them.

1.3 Organizational Levels of Life

An important task in scientific studies is to look for patterns in observations or data about the entities or processes under study. As a physicist would tell us, there is a great pattern of organization in nature starting at the atomic and even subatomic particle level. At a very fundamental level, all living organisms, including humans, are made from the same set of about 100 atoms from which any other material thing in our universe is made of. Multiple atoms bond together through some kinds of electromagnetic interaction to make molecules. For example, two atoms of oxygen bond to a hydrogen atom to make a molecule of water. Another example of a molecule is chlorophyll, a pigment molecule that is responsible for the green color of plant leaves. Millions of chlorophyll molecules are organized in a cell to convert the sunlight into carbohydrates in a process called *photosynthesis*. Life arises at a cellular level, and cells are made of molecules and atoms.

Life has hierarchical levels of organization (Figure 1.2) with an increasing complexity not found in nonliving entities. For all life, this organization begins with cells and goes beyond organisms. For a multicellular organism, there are levels of increasing complexity of life inside it beginning with a cell, the most fundamental structural and functional unit of life. Therefore, at its most fundamental level, a living organism is composed of one or more cells. In a multicellular organism, a set of cells organize into an array in order to perform a common function, and this array is called *tissue*. Tissues, in turn, organize into a structural unit called an *organ* such as a heart, which performs one or more specific tasks. A number of organs interacting to accomplish one or ore common tasks form an *organ system,* such as the digestion system. A multicellular organism is composed of multiple organ systems. A *system*, in general, is a combination of components that work together to accomplish a common task.

> **Caution!** Although cells are made of molecules and atoms, and molecules are made of atoms, life only begins with cells.

However, the organizational levels of life do not stop with an organism. After all, organisms interact with their environment, including other organisms. A group (populations) of organisms whose members have the potential of interbreeding and producing fertile and viable offspring is called a species. By definition, a species does not have the potential of interbreeding and producing viable offspring with another species. Furthermore, a species may be geographically distributed into subgroups called populations. In other words, a group of organisms of a given species living in an area at the same time is called a *population*. All populations of all species living in a specified area are called a *community*. These populations live close enough so that they have the potential of interacting with one another. One or more communities interacting with their physical environment makes what is called an *ecosystem*. The sum of all ecosystems on our planet is called the *biosphere*. This covers all parts of the Earth, environment included, where organisms live. Even though life begins with a cell, it is important to remember that a cell is made of molecules, which in turn are made of atoms, and that complex cellular functions can be truly understood only at molecular, atomic, and particle (subatomic) levels.

These different levels are not merely theoretical definitions; they represent the properties and interactions of life in very practical ways. Structure (anatomy) and functions (physiology)

Defining Life

of an organism are correlated; structure must support function for the organism to succeed, that is, survive and reproduce. This is expressed in a well-known biological principle: form fits function.

In Sections 1.2 and 1.3, we have explored the main characteristics of life, each of which represents one or more themes.

1.4 Themes of Biology

The following is a list of themes in biology, the science of life, most of which we have already mentioned while exploring the characteristics of life:

- Cells are the fundamental structural and functional units of any living organism.

- The continuity of life from one generation to the next is based on DNA, which is transferred from parents to their offspring.

- New properties emerge at each level of the complexity of life and are called emergent properties. These properties are due to the arrangement of and interaction between the parts of the system and do not belong to any specific part.

- Structure and function are correlated at all levels of complexity in biological organization: form fits function.

- Organisms interact with their environment, exchanging nutrients and energy.

- Feedback mechanisms contribute to regulating important life processes such as growth and reproduction and to maintaining homeostasis. Energy and nutrients are required to support these feedback mechanisms.

- Evolution is the central theme of life, which explains the unity behind the diversity of life.

Molecular and Cellular Biology

Evolution, the unifying theme of life, helps to organize incredibly diverse life forms into groups such as species, genres, kingdoms, and domains.

1.5 Organizing Diversity of Life into Domains

In order to make sense of diversity, scientists have been classifying organisms into various groups based on structural, functional, and other features. According to the classification scheme developed by a Swedish physician named Carolus Linnaeus (1707-1778), organisms at the highest level are classified into five kingdoms: Monera, Protista, Fungi, Plantae, and Animalia. This classification is largely based on cellular organization and modes of nutrition of organisms. Cellular organization is based on whether the organisms have eukaryotic cells or prokaryotic cells and whether the organisms are unicellular or multicellular. As illustrated in Figure 1.3, all prokaryotes are classified into the Kingdom Monera, whereas eukaryotes occupy the other four kingdoms. The Kingdom Protista and the Kingdom Monera are composed of mostly unicellular organisms, whereas the other kingdoms are mostly multicellular. Kingdom Plantae is made up of autotrophs, whereas the Kingdom Fungi and the Kingdom Animalia are heterotrophs.

> **Note.** There are two types of cells: *eukaryotic cells*, which have membrane-bound internal structures and *prokaryotic cells*, which do not have membrane-bound internal structures.

Based on relatively recent molecular data, these kingdoms have been organized into three domains, as illustrated in Figure 1.3. These domains are described as follows:

Defining Life

Domain Bacteria. This domain consists of most diverse and widespread unicellular organisms made of prokaryotic cells. These organisms are a subset of the Kingdom Monera.

Domain Archaea. This domain consists of prokaryotes, including those that live in extreme environments such as salty lakes and boiling hot springs. Domains Archaea and Bacteria combine to form Kingdom Monera.

Domain Eukarya. This domain consists of all organisms made of eukaryotic cells. This domain includes four kingdoms: Protista, Fungi, Plantae, and Animalia.

These three domains make up a field studied by biology, a branch of science. Science is the study of nature. However, it is not just any study of nature, but a study performed by using the scientific method.

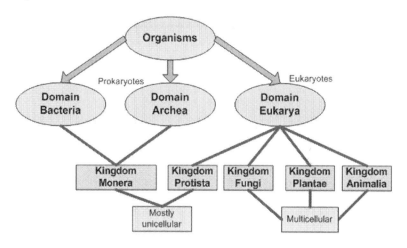

Figure 1.3 Classification of the diversity of life.

1.6 Science and Scientific Method

Science is a way to obtain knowledge about natural entities and phenomena. The central tool of science is scientific inquiry. Two main types of scientific inquiries are discovery science and hypothesis-based science.

Discovery Science. Discovery science can be defined by its output, which is the description of entities, structures of entities, and phenomena. This output is produced through observations and data analysis. It is also called empirical science. For example, determining the DNA sequences of different species is a discovery science. One problem (or limit) of discovery science is that you may produce an overwhelming amount of data without any central principle that unifies it all. For example, these days we have tons of data available from genetic studies such as DNA sequencing and genomics. However, this data may be used to discover fundamental principles of nature by using hypothesis-based science.

> **Hypothesis.** A *hypothesis* is a tentative description or explanation of specific observables about a physical entity or phenomenon or a set of entities or phenomena.

Hypothesis-Based Science. Whereas discovery science is about describing nature, hypothesis-based science is mostly about explaining nature by exploring the root causes of observations or data. Therefore, one can argue that hypothesis-based science is more fundamental than discovery science. The theory of evolution and cell theory are examples of hypothesis-based science. Hypothesis-based science follows the scientific method (Figure 1.4).

Defining Life

The scientific method is a technique or a process used for acquiring knowledge through verification. The main steps of this technique are as follows:

1. Make observations and state the problem. For example, collect data and ask a question based on patterns in the data.

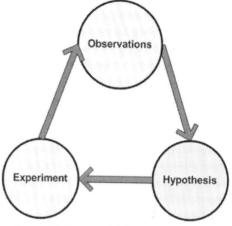

Figure 1.4 Three crucial elements of the scientific method.

2. As a solution to the problem or an answer to a question, propose a falsifiable and testable hypothesis. Make predictions based on the hypothesis.

3. Perform experiments to test your hypothesis by testing predictions based on the hypothesis. If the experimental results are in agreement with the predictions, then the experiment supports the hypothesis. Go to Step 4. If the experimental results are in disagreement with the predictions, then the experiment falsifies the hypothesis. In this case, depending on the issue at hand, you may proceed to Step 4 or go back to Step 1.

4. Share your results with the scientific community and provide enough details so that other scientists could repeat the experiment by reading your report.

If your hypothesis was falsified by the data, you can modify or change your hypothesis and go through these steps again.

Molecular and Cellular Biology

A hypothesis or a set of hypotheses develop into a theory only after standing years of scientific scrutiny and tests. In science, when we call something a theory that means it is well tested and that it has withstood all (or most) of the scientific tests performed so far. Theories are continuously tested and modified or improved if new information or data demands it. Thus, a theory usually has a much broader scope and is more general than a hypothesis. As a result, a theory also has a greater body of evidence that supports it than a hypothesis.

1.7 In a Nutshell

Life begins at the cellular level; that is, cells are the fundamental functional and structural units of life. Historically too, life began with a cell or a set of the same or similar cells and evolved into a diverse spectrum of organisms through descent with modification from the same ancestor. This theme of evolution explains the unity behind the diversity of life and all other major characteristics of life or themes of biology, some of which are listed below:

- At each organizational level of life, the sum is greater than its parts; that is, new properties called emergent properties appear, which do not belong to any specific part.
- Organisms grow and reproduce.
- Organisms store information necessary for life processes in DNA molecules and transfer it to their offspring.
- Organisms use energy and nutrients to maintain homeostasis, grow, and reproduce.
- Organisms interact with their environment in many ways, for example, to obtain energy and nutrients.
- Organisms respond to their environment.

1.8 Review Questions

1. Which of the following is an example of the level of organization from which life begins?

    A. reproduction system in a human

    B. human

    C. blood cell

    D. heart

    E. muscle tissue

    F. DNA

2. What is true about evolution?

    A. Chimpanzees are the ancestors of humans.

    B. Different species evolved by descending with modification from a common ancestor.

    C. Humans evolved from monkeys.

    D. A. and B.

3. Which of the following is not true?

    A. Prokaryotes are split into two domains: Archaea and Bacteria.

    B. Kingdom Protista consists of mostly unicellular organisms.

    C. Fungi are mostly multicellular.

    D. All prokaryotes are mostly unicellular, and all unicellular organisms are prokaryotes.

Molecular and Cellular Biology

4. Prokaryotes belong to _____ kingdom and _____ domain(s).
   A. Monera, Archaea and Bacteria
   B. Monera, Bacteria
   C. Animalia, Archaea and Bacteria
   D. Protista, Bacteria

5. Evolution explains _____.
   A. the unity behind the diversity of life
   B. the growth and metabolism of organisms
   C. the response of organisms to the environment
   D. all of the above

6. Which kingdom consists of mostly unicellular eukaryotes?
   A. Monera
   B. Prokarya
   C. Protista
   D. Animalia

7. What is true about a scientific theory?
   A. It is an untested speculation.
   B. It has a broader scope than a hypothesis.
   C. It is more general than a hypothesis.

D. It has a much larger body of evidence that supports it than a hypothesis.

E. All of the above.

F. B., C., and D.

8. Which of the following characteristics is not unique to life and applies to both living and nonliving entities?

    A. development and growth

    B. reproduction

    C. response to the environment

    D. structure

9. Classification of organisms into five kingdoms is based on _____.

    A. the number of cells and mode of nutrition

    B. the mode of nutrition and evolutionary history

    C. the size and shape of the organisms

    D. evolutionary history and genetic analysis

10. Classification of organisms into three domains (Bacteria, Archaea, and Eukarya) is based on _____.

    A. the number of cells and mode of nutrition

    B. the mode of nutrition and evolutionary history

    C. the size and shape of the organisms

    D. evolutionary history and genetic analysis

Molecular and Cellular Biology

11. Which of the following is one of the differences between theories and hypotheses?

    A. Theories are proved true, and hypotheses are untested guesses.

    B. A theory is a hypothesis that has been tested to be true.

    C. A theory is a combination of many hypotheses.

    D. A theory has a wider scope than that of a hypothesis.

1.9 Answer Key

1. C.
2. B.
3. D.
4. A.
5. D.

6. A.
7. F.
8. D.
9. A.
10. D.
11. D.

Notes:

Q5. Evolution is the unifying core theme in biology that links together all the other major themes.

Q9. The number of cells are part of cellular organization.

Chapter 2

Chemistry of Life

2.1 Biochemistry: The Big Picture

Biochemistry is the study of chemical processes in living entities, including organisms and any living matter. Cells, the basic units of life, are composed of molecules, which in turn are composed of chemical elements (atoms). Some elements and molecules inside the cell interact and chemically react with one another. A cell is considered to be a self-sustained chemical factory in which thousands of different chemical reactions occur and interact smoothly to maintain life. These reactions are interconnected, for example, to make what are called metabolic pathways (Figure 2.1), which involve the regulation and interaction of these reactions (The labels in this figure may not be readable; it is included here just to give you a feel for the enormity of the pathways). This system of chemical reactions including pathways and regulations has been transferred from one generation to the next for millions and millions of years while going through evolution at the same time.

In this chapter, we review basic biochemistry by exploring very basic concepts of chemistry relevant to living systems. The chemical connection of life begins with elements of the periodic table from which all matter is composed.

Molecular and Cellular Biology, by Paul Sanghera
Copyright © 2015 Infonential.

Molecular and Cellular Biology

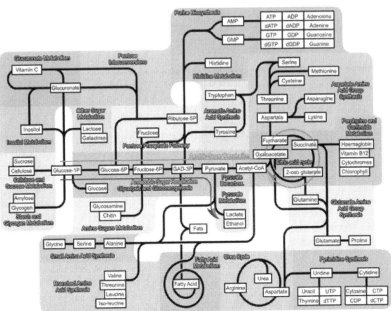

Figure 2.1 Major metabolic pathways. A metabolic pathway is a series of interconnected chemical reactions that either composes a complex molecule or breaks down a complex molecule to simpler ones.

2.2 Chemical Connection of Life

All living and nonliving entities are matter. Matter, living and nonliving, is anything that occupies space. All matter is composed of atoms, which in turn are composed of subatomic particles such as protons, neutrons, and electrons. An element is a substance that is composed of the same type of atoms and cannot be broken down to other substances by chemical reactions. A compound is a substance that is composed of two or more elements in a fixed ratio. For example, hydrogen (H) is an atom and an element, whereas water is a compound. The smallest unit of water is composed of two hydrogen (H) atoms and one oxygen (O) atom, and it is called the water molecule (H_2O). A molecule is a chemical combination of two or more atoms. Two suitable molecules can combine to make a more complex molecule. For example, a large number of amino acid molecules chemically combine to make protein molecules.

Table 2.1 Main elements that naturally occur inside a human body.

Element	Symbol	Atomic Number	Percentage of Human Body Mass
Oxygen	O	8	65.0
Carbon	C	6	18.5
Hydrogen	H	1	9.5
Nitrogen	N	7	3.3
Calcium	Ca	20	1.5
Phosphorus	P	15	1.0
Potassium	K	19	0.35
Sulfur	S	16	0.25
Sodium	Na	11	0.15
Chlorine	Cl	17	0.15
Magnesium	Mg	12	0.05

Molecular and Cellular Biology

As you learned in Chapter 1, a cell, the basic unit of life, consists of atoms and molecules. Before we explore how atoms combine to make molecules, it is essential to review some basic concepts of chemistry.

2.3 Basic Concepts

Here are the definitions of some basic concepts that you need to know to study the chemical aspects of biological systems.

Atomic number. The atomic number of an element is the number of protons in each atom of the element. The number of electrons in a neutral atom is also equal to its atomic number. For example, atomic number of hydrogen is 1.

Mass number. The mass number of an element is equal to the sum of the number of protons and the number of neutrons in each of its atoms. For example, mass number of hydrogen is also 1 because its atom contains no neutron.

Atomic mass. The total mass of an atom is usually measured in atomic mass unit (amu), which is also called dalton. Both protons and neutrons weigh about 1 dalton each, and because an electron weighs only about 1/2000 of a proton, the atomic mass of an element is about equal to its mass number in daltons. For example, the mass number of sodium (Na) is 23, and its atomic mass is about 23 daltons.

Isotopes. The isotopes of an element are its different atomic forms with the same number of protons but a different number of neutrons, which give rise to different mass numbers and hence different atomic masses for different isotopes. Table 2.2 presents some isotopes of carbon as an example. An isotope symbolically is represented by the symbol of the atom with its atomic number as its subscript and its mass number as its superscript. For example, the symbol $^{12}_{6}C$ represents a carbon isotope with a mass number of 12 and, of course, an atomic number of 6.

Table 2.2 Some isotopes of carbon.

Isotope	Atomic Number	Number of Protons	Mass Number	Number of Neutrons
$^{12}_{6}C$	6	6	12	6
$^{13}_{6}C$	6	6	13	7
$^{14}_{6}C$	6	6	14	8

All different isotopes of the same element behave identically in chemical reactions, but they differ in their nuclear properties. For example, some isotopes are more prone to radioactive decay than others, and this understanding is used in some biological research and applications. For instance, some radioactive elements are used to determine the date (age) of ancient objects such as fossils.

Atomic shells. An atomic shell is a region around the atom in which electrons can have a certain amount of energy called the energy level or principle energy level. An atom has, in principle, infinite numbers of energy levels. Each energy level (shell), however, can have only a finite number of electrons. The maximum number of electrons in the nth shell is given by $2n^2$. For example, the number of maximum electrons in the 1st shell (ground state) of an atom is equal to

Molecular and Cellular Biology

2, and the maximum number of electrons in the next higher shell (n=2) is equal to 8.

Atomic orbitals. The electrons in a shell are actually held in certain regions in the shell called orbitals. Each orbital can hold a maximum of 2 electrons. Of course, the electrons move around in their orbitals.

Valence shell. The outermost shell occupied by electrons in an atom is called the valence shell.

Valence electrons. The electrons in a valence shell are called valence electrons. These electrons are the easiest ones to knock out of the atom because they are more loosely bound to the nucleus of the atom than the electrons in the inner shells, which are closer to the nucleus. As a result, it is the valence electrons that participate in chemical reactions and largely determine the chemical properties of the elements.

Polar molecules. As you have just learned, atoms in a covalent bond share electrons. If the electrons are shared equally, the bonds are called nonpolar. For example, when two atoms of the same type bond such as those in chlorine (Cl_2), the resulting bond is nonpolar. When one of the atoms in a bond attracts the shared electrons more strongly than the other atom, the resulting bond is called polar. If the difference in the attractive forces of the two atoms for the shared electrons becomes large enough, the bond becomes ionic. For example, the bond between sodium (Na) and chlorine (Cl) atoms in a sodium chloride (table salt) molecule is an ionic bond. The ability of an atom in a bond or a molecule to attract the electrons to itself is called *electronegativity*. In a polar bond, there is a concentration of negative charge on the atom with greater electronegativity and a concentration of positive charge on the atom with less negativity. These two concentrations of positive and negative charges give rise to an arrangement called a dipole in physics. This internal imbalance of charge gives rise to a polarity whose strength is measured with a quantity called a *dipole moment*. A molecule with a net nonzero polarity or dipole moment is called a polar molecule. An example of a polar molecule is water (Figure 2.4 in Section 2.5), which is discussed further on in this chapter.

Polar molecules of one kind mix with polar molecules of another kind to make solutions. Two or more different kinds of nonpolar molecules also mix together to make solutions. What are solutions?

Solution. A solution is a homogeneous mixture of two or more substances. A solvent is the substance that acts as a dissolving medium for the solution; it is usually the component of the solution that is present in a greater amount. A solute is a substance that dissolves in the solvent; it is usually the component of the solution that is present in a smaller amount. Water is an example of a solvent, and sugar is an example of a solute. One of the quantities used to measure the concentration of the solution is called *molarity*, and it is symbolized by M, which is the number of moles of solute in one liter of solution.

Now that you know the very basic concepts of chemistry, we can explore how atoms combine (bond) to make molecules.

2.4 Atomic Bonds

An atomic bond is the binding of two similar or different types of atoms in a stable arrangement. Bonding happens naturally as it leads to a system with lowered energy and therefore increased stability. The structure that results from the bonds of atoms from two or more different elements is called a molecule. If the molecules contain more than one type of atoms, the substance made of these molecules is called a molecular *compound*.

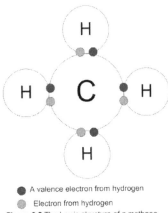

Figure 2.2 The Lewis structure of a methane molecule showing four covalent bonds between a carbon atom and four hydrogen atoms.

Bonds are of two types: covalent bonds and ionic bonds.

Covalent bonds. A covalent bond between two atoms is a bond in which the atoms share two electrons. A covalent bond is a two electron bond. The smallest unit of a substance composed of two or more atoms bonded together with a covalent bonds is called a molecule. Figure 2.2 illustrates the molecular structure of a compound called methane, which is composed of four covalent bonds made by a carbon atom with four hydrogen atoms.

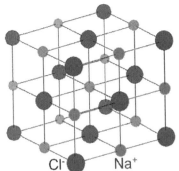

Ionic bonds. An ionic bond between two atoms is a bond in which one atom loses one or more electrons and the other atom gains these electrons. The atom that loses electrons becomes positively charged and is called a *cation*, and the atom

Figure 2.3 Cubic structure of a sodium chloride crystal.

that gains those electrons becomes negatively charged and is called an *anion*. The ionic bonding occurs due to the electrostatic force between the cation and the anion. For example, a sodium (Na) atom loses an electron to become a cation (Na^+), and a chlorine atom (Cl) loses an electron to become an anion (Cl^-), and the two join together in an ionic bond to form sodium chloride (NaCl), which is table salt. The structure of a table salt crystal is illustrated in Figure 2.3. Ionic compounds such as sodium chloride form extended crystal lattices to maximize the positive and negative electrostatic interactions in order to gain stability. Substances or compounds made of ions are called ionic compounds.

> **Note.** Most matter or substances are made of molecules or ions, both of which in turn are made of atoms.

Chemical reactions occur by breaking existing chemical bonds and making new ones, which may result in consuming some molecules (reactants) and creating some new molecules (products).

Molecules of water, made from covalent bonds between hydrogen and oxygen, played a very significant role in the survival and hence the evolution of life on Earth.

2.5 Water: The Universal Medium of Life

Life on Earth is highly dependent on water. For example, cells, the fundamental structural and functional units of life, are about 70 to 95% water. The most salient characteristics of water that support life emerge from a single fact: the ability of water molecules to form hydrogen bonds with other water molecules. This ability of water arises from the fact that the water molecule is a polar molecule. A hydrogen bond is a bond that is formed by an attractive intermolecular interaction that occurs when a hydrogen atom bonded to an F, N, or O atom is electrostatically attracted to a lone pair of electrons on another F, N, or O atom. Figure 2.4 illustrates these bonds between water molecules.

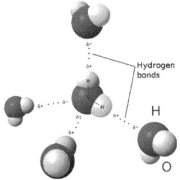

Figure 2.4 Hydrogen bonds between water molecules

The following four properties of water emerging from hydrogen bonds contribute heavily to the suitability of Earth for life:

Cohesion. Due to hydrogen bonding, water molecules stick together. This property is called *cohesion*, and it facilitates processes such as the flow of water from the roots to the leaves of the plants.

Ability to moderate temperature. Water absorbs heat to form hydrogen bonds and releases heat when hydrogen bonds break. This gives water the capacity to absorb or release considerable heat without much change in its temperature. In other words, water has a high value of *specific heat*, the quantity of heat absorbed or released by one gram of water to change its temperature by 1°C. Thus, water moderates air temperature by absorbing heat from the warmer air and releasing it to the colder air with only a slight change in its temperature. This is why the temperature in a coastal area is very moderate and gives rise to a milder climate as compared to the temperature of inland areas. In addition, because organisms consist of a large amount of water, they are able to maintain their body temperatures essentially constant even when going through much hotter and colder environments.

Expansion upon freezing. When water is liquid, hydrogen bonds are constantly forming and breaking. When water freezes into ice, the water molecules get locked into a crystalline lattice due to the hydrogen bonds that keeps the water molecules farther away from each other as compared to water in liquid form. This is why ice is less dense than water and hence floats on water. The layer of ice on lakes produced by the expansion of water upon freezing actually keeps the water underneath the layer from freezing and protects life in the water.

Versatility as a solvent. Water is a very versatile solvent as it dissolves many other substances. To support life, we need solvents, and nothing works better than water as a solvent. This property of water also emerges from the polarity of its molecules. Water dissolves all polar and ionic substances and also the polar parts of nonpolar substances. For example,

Molecular and Cellular Biology

some proteins such as some types of hormones dissolve in water due to the polar parts of their molecules.

> **Note.** Substances that have an affinity for water and easily dissolve in it are called hydrophilic, while substances that do not have an affinity for water and resist dissolving in water are called hydrophobic.

Water is the universal medium of life, whereas the carbon atom is the basis of life.

2.6 Carbon: The Basis of Life

Carbon is often said to be the backbone of life on Earth. Carbon occupies a whole branch of chemistry called organic chemistry, the study of organic compounds, because all organic compounds contain carbon. The molecules of life are organic molecules. Plants use the energy from sunlight to transform atmospheric CO_2 into the molecules of life such as carbohydrates. Animals, including humans, feed on plants and use these molecules as raw material to make their own molecules of life in their cells.

Figure 2.5 The Lewis structure of a carbon atom showing the four valence electrons.

A carbon atom has four valence bonds, and this is the fundamental characteristic that gives carbon a set of uncommon emergent properties that enables it to be the backbone of life. These properties include the following:

- Due to its four valence electrons (Figure 2.5), a carbon atom can form four covalent bonds with up to four atoms.
- Carbon can form covalent bonds with other carbon atoms and also with some noncarbon atoms such as hydrogen

(H), oxygen (O), nitrogen (N), sulfur (S), and phosphorus (P).

- The four covalent bonds may come in any combination of single or double bonds.
- Carbon has diverse bonding capabilities. While bonding with one another, carbon atoms may form different shapes such as linear chains of various lengths, linear chains with branches, rings of various sizes, and rings with branches. This gives rise to carbon skeletons of various shapes and sizes.
- The ability of carbon to bond to noncarbon atoms enables the carbon skeletons to be attached to various chemical groups called functional groups, which participate in chemical reactions and offer diversity in the molecules that can be made on top of a carbon backbone. These functional groups are keys to the functions of various biological molecules.
- The molecules of life (carbohydrates, nucleic acids, lipids, and proteins), which will be discussed in the next chapter, are basically composed of carbon atoms bonded to one another and to noncarbon atoms.

In a nutshell, carbon skeletons facilitate or give rise to single bonds, double bonds, different chain lengths, ring formation, branching, and attachment of various functional groups attached to carbon atoms. Carbon skeletons also produce or support three types of isomers: structural, geometric, and enantiomers. All this gives rise to the diversity of molecules based on these carbon skeletons. The underlying source of this diversity remains the four valence electrons of the carbon atom.

But what are isomers? Let us explore.

2.7 Isomers

Isomers are compounds that have the same molecular formula but different structures. How do they differ in structure? The answer to this question gives rise to three main types of isomers:

Structural isomers. Also called constitutional isomers, these isomers differ in the covalent partners of the carbon atoms in the chain and also in the location of double bonds.

Geometric isomers. These isomers differ in the arrangement of atoms around a double bond.

Enantiomers. These isomers differ in the spatial arrangement of atoms around an asymmetric carbon atom, giving rise to molecules that are mirror images of each other.

In Section 2.6, we mentioned that functional groups contribute to the diversity of organic molecules that can be formed with carbon skeletons. But what really are functional groups? We will explore them next.

2.8 Functional Groups

A functional group is a group of atoms within an organic molecule that acts as a site of chemical reactivity and hence determines how the molecule behaves or functions. Each type of functional group undergoes the same kinds of chemical reactions in all different molecules. Some common functional groups in organic compounds are presented in Table 2.3.

Here are a few notes to explain some of the terms in the table:

- Two sulfhydryl groups can make an S-S covalent bond called a *disulfide bridge* that helps stabilize protein structure.
- An *aldehyde* is a compound in which the *carbonyl group* (C=O) is at the end of the carbon skeleton.
- A *ketone* is any compound in which the carbonyl group is within the carbon skeleton and not at the end.

Molecular and Cellular Biology

Table 2.3 Some common functional groups in organic compounds.

Functional Group	Structure	Compound	Functional Properties	Example
Amino	H–N–H	Amines	Water soluble Weak base	Valine (amino acid)
Carbonyl	–C(=O)–	Aldehydes Ketones	Water soluble	Valine
Carboxyl	–C(=O)–OH	Carboxylic acids	Water soluble Polar Acidic	Acetic acid
Hydroxyl	–OH	Alcohols	Water soluble	Methanol

Methyl	—CH$_3$	Methylated compounds	Insoluble in water	Fatty acids (Saturated / Unsaturated)
Phosphate	—O—P(=O)(O⁻)—O⁻	Organic phosphates	Water soluble, Acidic	Nucleotides

Sulfhydryl	—SH	Thiols	disulfide bridge	Methane Thiol	Ethane Thiol

2.9 Acids and Bases

An acid is a substance that when added to a solution increases the hydrogen ion concentration of the solution. The concentration of acid in a solution is called acidity; the higher the concentration of hydrogen ions in the solution, the greater is the acidity value of the solution. It is crucial to maintain the acidity value within a very narrow range for living organisms to survive. Homeostasis discussed in Chapter 1 accomplishes this task.

Acidity is measured by a variable written as pH, which is defined by the following formula:

$$pH = -\log_{10}[H^+] \qquad (2.1)$$

In Equation 2.1, $[H^+]$ represents the concentration of hydrogen ions in the solution measured in molarity, M, and brackets mean concentration. Note that due to the negative logarithm, the higher the concentration of hydrogen ions, the lower is the pH value. This means that a lower pH value represents higher acidity. The pH value of a solution varies between 0 and 14. A neutral solution has a pH value of 7, which is the pH value of water. Acidic solutions have a pH value lower than 7, whereas basic solutions have a pH value greater than 7. The pH value of most fluids inside living organisms is between 6 and 8 with a few exceptions such as the digestive juice of the human stomach, which has a pH of 2.

> **Fascinating Fact!**
>
> The digestive juice in your stomach, which is a HCl solution, is secreted by the gastric glands. These glands get to work as soon as you begin tasting or chewing (ingesting) food. They get to work even when you just smell, see, or think of food. The secretion continues as long as the stomach has a pH of 2, which is necessary to activate a protein (enzyme) called pepsin that participates in digesting the food by breaking down proteins. Your gastric glands may secrete as much as 1.4 L of this acidic digestive juice per day.

Contrary to an acid, a base is a substance that when added to a solution decreases the hydrogen ion concentration of the solution. A base decreases hydrogen ions by producing OH⁻ ions, which react with H⁺ ions and hence consume them. The *basicity* of a solution is measured by a variable written as pOH, and it is defined by the following formula:

$$\text{pOH} = -\log_{10}[OH^-] \quad (2.2)$$

In Equation 2.2, $[OH^-]$ represents the concentration of hydroxyl ions in the solution. It turns out that the product (multiplication) of the concentrations of hydrogen ions and hydroxyl ions is always equal to a constant expressed by the following equation:

$$[H^+][OH^-] = 10^{-14} \quad (2.3)$$

Taking the log of both sides of Equation 2.3 and using Equation 2.2, one can write:

$$pH + pOH = 14$$

A buffer is a substance that resists the change in pH in a solution. For example, carbonic acid (H_2CO_3) acts as a buffer in human blood. It forms when carbon dioxide (CO_2) reacts with water (H_2O) in blood plasma, as shown in the following reaction:

$$H_2O + CO_2 \rightarrow H_2CO_3 \quad (2.4)$$

Where does CO_2 come from? We breathe in oxygen (O_2), which is an input to the cellular metabolism, and this metabolic process generates CO_2 as one of its end products. While we breathe out some CO_2, the rest dissolves in body fluids such as blood plasma and saliva. Carbonic acid resists the change in pH by disassociating a hydrogen ion and a bicarbonate ion (HCO_3^-) and associating back to carbonic acid as required to keep pH stable. This process is represented by the following chemical reaction:

$$H_2CO_3 \rightleftharpoons H^+ + HCO_3^- \quad (2.5)$$

The pH value is kept stable by establishing the equilibrium of this chemical reaction, which shifts to the right to resist the increase in pH by producing more hydrogen ions and then shifts to the left to resist the decrease in pH by consuming hydrogen ions. Shifting to the right means producing more substances on the right of the arrows, and shifting to the left means producing more substances on the left of the arrows.

> **Fascinating Fact!**
>
> The normal pH in our arterial blood stays in a very narrow range from 7.35 to 7.45. If the pH value of our arterial blood drops below 6.8 or rises above 8.0, death may result because our cells cannot function properly under these pH conditions.

2.10 In a Nutshell

An understanding of biochemistry is essential to understanding life. Here are some basic points:

- Molecules of life are composed of atoms from the periodic table of elements.
- All different isotopes of the same element behave identically in chemical reactions, but they are different in their nuclear properties.
- The chemical properties of elements are largely determined by the number of valence electrons they have.
- Elements in the same row of the periodic table have the same number of electron shells in their atoms, and elements in the same column have the same number of valence electrons in their atoms.

Molecular and Cellular Biology

- Covalent bonds form molecules and molecular compounds and substances, whereas ionic bonds form ionic compounds, which are usually crystals with atoms arranged in a three-dimensional lattice.
- Chemical reactions occur by breaking existing chemical bonds and making new ones.
- The proportions of major elements such as carbon, oxygen, hydrogen, and nitrogen are very uniform across all living organisms.
- Four valence electrons in the carbon atom enable it to form diverse molecules.
- Proper acidic and basic conditions such as pH value inside living organisms are critical to maintain life.
- The properties of water play a significant role in maintaining life on Earth.

Biochemistry includes the study of the formation, structure, function, interactions, and breakdown of the molecules of life that make a cell. These biomolecules are discussed in the next chapter.

2.11 Review Questions

1. Your body, just like the body of many other mammals, regulates body temperature (keeps body temperature low) in hot weather by losing heat energy through the evaporation process of sweating. What in water directly facilitates the lowering of your body temperature in this process?

    A. breaking of hydrogen bonds

    B. making of hydrogen bonds

    C. breaking of polar covalent bonds

D. making of polar covalent bonds

2. Which of the following is hydrophobic?
   A. salt
   B. sugar
   C. vegetable oil
   D. wheat flour
   E. paper

3. A water molecule has ____ covalent bond(s), and a methane molecule has ____ covalent bond(s).
   A. 1...1
   B. 1...2
   C. 2...4
   D. 2...2

4. A water molecule makes ____ hydrogen bond(s) with other water molecules.

Molecular and Cellular Biology

   A. 1
   B. 2
   C. 3
   D. 4

5. A carbon atom can form up to how many covalent bonds?

   A. 1
   B. 2
   C. 3
   D. 4

6. Your body takes in carbon atoms by ingesting organic molecules in food. Which of the following is the correct statement about these carbon atoms?

   A. These carbon atoms were introduced into organic molecules by plants.
   B. You can get these carbon atoms by eating meat as well.
   C. These carbon atoms ultimately came from the atmosphere.
   D. These carbon atoms ultimately came from carbon dioxide.
   E. All of the above.

7. Carbonic acid in our blood plasma contributes to homeostasis by _____.

A. reducing the value of pH
B. elevating the value of pH
C. acting as a buffer
D. producing hydrogen ions

8. Which of the following correctly represents the acidity of the fluid in your stomach?

   A. pOH=12
   B. pOH=2
   C. $[H^+]$=2 M
   D. $[OH^-]$=10^{-12}M
   E. A. and D.

9. Which of the following is a true statement about functional groups?

   A. They determine the chemical reactivity of the molecule of which they are a part.

Molecular and Cellular Biology

B. They contribute to the diversity of molecules that can be made from a carbon skeleton.

C. They are all polar.

D. They are all nonpolar.

E. A. and B.

10. You have two drinks in front of you: a can of Coke with a pH of 2.7 and a glass of orange juice with a pH of 3.5. Which of the following is a true statement about these drinks?

   A. Orange juice is more acidic than Coke.

   B. Coke is more acidic than orange juice.

   C. Orange juice is more basic than Coke.

   D. B. and C.

2.12 Answer Key

1. A.
2. C.
3. C.
4. D.
5. D.
6. E.
7. C.
8. E.
9. E.
10. D.

Notes:

Q6. Plants took these carbon atoms from the atmosphere in the form of carbon dioxide and incorporated them into the organic molecules that they made by using photosynthesis.

Q8. pH of stomach = 2 ➔ pOH = 14 – 2 = 12 = - $\log_{10}[OH^-]$

➔ $[OH^-] = 10^{-12}$

Molecular and Cellular Biology

Chapter 3

Molecules of Life

3.1 Molecules of Life: The Big Picture

Cells, the fundamental units of life, are filled with and run by small and large molecules. The most crucial of these molecules are called the molecules of life, which are made in the living cells of all organisms. These molecules can be grouped into the following four categories: carbohydrates, lipids, proteins, and nucleic acids. Carbohydrates are sugars and polymers of sugars, and they mostly function as a source of energy. Some carbohydrates also play structural roles. Lipids are oily or waxy molecules, which are key components of cell membranes and are also used to store energy. Proteins are polymers of amino acids folded into appropriate shapes, and they make components of many parts of an organism's body and are responsible in one way or another for almost all dynamic functions of the body. Proteins are synthesized in a cell according to the instructions chemically coded in genes, which are housed in DNA molecules. DNA molecules are nucleic acids, and they are the genetic material inherited by organisms from their parents.

Due to their size, carbohydrates, proteins, and nucleic acids are also called macromolecules. Lipids can be large, but most of them are not large enough to be called macromolecules. All of these molecules are carbon-based, and their chemistry is included in the field called organic chemistry. The study of

Molecular and Cellular Biology, by Paul Sanghera
Copyright © 2015 Infonential.

Molecular and Cellular Biology

these molecules of life and the related processes make up the field of biochemistry.

In this chapter, we explore the basics of these molecules of life beginning with carbohydrates.

3.2 Carbohydrates

Carbohydrates are sugars and polymers of sugars made of carbon, hydrogen, and oxygen. The general molecular formula for sugars is $C_nH_{2n}O_n$, where n is an integer greater than or equal to 1. Almost all organisms use carbohydrates as a source of energy. They also serve as structural material. Carbohydrates can be grouped into the following three categories:

Monosaccharides. These are the simplest carbohydrates with one monomer, the simplest sugars used for energy. For example, glucose, the most popular monosaccharide with the molecular formula $C_6H_{12}O_6$, is a major nutrient for cells. It is the main product of photosynthesis in plants. Fructose, the sugar in fruits, and galactose, a sugar in milk and yogurt, are other examples of monosaccharides.

Disaccharides. Disaccharides are carbohydrates that are composed of two monomers. They are also used for energy. Some examples of disaccharides are maltose, a sugar found in germinating seeds such as barley; lactose, the sugar present in milk; and sucrose, which is table sugar.

Oligosaccharides. Oligosaccharides are carbohydrates that are composed of a few monomers. These are used for cell-to-cell recognition, communication, immune cell functioning, and cell trafficking. Examples of oligosaccharides are fructo-oligosaccharides and polymers of fructose monomers, which are found in many vegetables.

Molecules of Life

Polysaccharides. Polysaccharides are complex carbohydrates containing many monomers. They are used for energy storage and building material for structures. Here are some examples of polysaccharides. A polysaccharide called *glycogen* is composed of thousands of glucose molecules. Humans and other vertebrates store energy (glucose) in the form of glycogen largely in their livers and muscle cells. It is the animal equivalent of *starch*, another polysaccharide, which stores energy (glucose) in plants in the form of starch granules (particles), and is largely stored in cellular structures called *plastids*. *Cellulose*, which strengthens the cell walls in plants, is an example of a structural carbohydrate; cotton, paper, and wood are largely cellulose or products of cellulose. *Chitin* is another example of a polysaccharide; it is actually a modified carbohydrate, which has nitrogen containing groups attached to its many-glucose monomers. It is used to give strength to hard parts of the body of some organisms. For example, the exoskeletons of arthropods such as crustaceans, insects, and spiders are largely made of chitin. In addition, fungi use chitin to develop their cell walls.

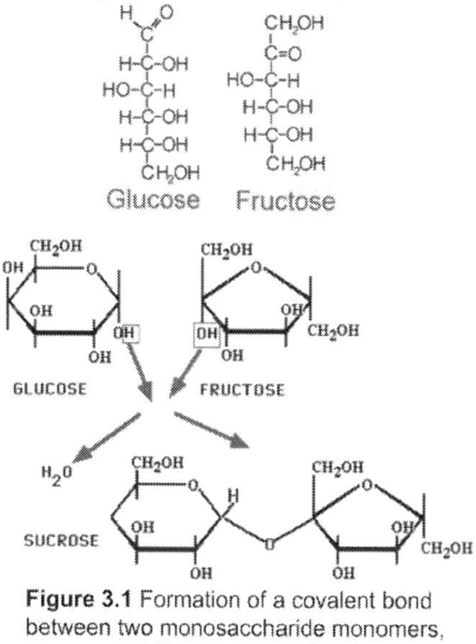

Carbohydrates are either made of just one unit called a monomer such as glucose, or they are polymers formed by combining several monomers. How are polymers formed from monomers? As illustrated in Figure 3.1, two monomers

Figure 3.1 Formation of a covalent bond between two monosaccharide monomers, glucose and fructose, through dehydration.

combine to form a covalent bond between them by releasing a water molecule in the process. This process is called *dehydration,* and the covalent bond or link made this way is called a *glycosidic link*. In this example, a glucose molecule and a fructose molecule, both monosaccharides, combine to form a sucrose molecule, a disaccharide. A maltose molecule forms by combining two glucose molecules, whereas a sucrose molecule forms by combining a glucose molecule with a galactose molecule. Dehydration is also used to make longer polymers by adding one monomer at a time. This process is not limited to carbohydrates; it is also used in making other molecules of life: lipids, proteins, and nucleic acids.

3.3 Lipids

Lipids are fatty, oily, or waxy organic compounds that dissolve poorly if at all in water. Just like carbohydrates, they are composed of carbon, hydrogen, and oxygen; however, the ratio of hydrogen atoms to carbon atoms in lipids is greater than 2:1. Lipids are large biological hydrophobic molecules, but they are not large enough to be considered macromolecules. They are not true polymers as they are not formed by joining large number of same or similar types of monomers. Fatty acids and an alcohol called glycerol are two important components of many lipids. A fatty acid is a carboxylic acid with a long hydrocarbon chain. There are many types of fatty acids distinguished by the length of their carbon chain and the number and location of double bonds.

Some of the important types of lipids are fats, phospholipids, and steroids.

Molecules of Life

Fats. Fats are simple lipids with one, two, or three fatty acids dangling like tails from a glycerol molecule. Figure 3.2 illustrates the formation of a triglyceride fat (triacylglycerol) by combining a glycerol molecule with three fatty acid molecules through dehydration. Simple lipids such as fats are used for insulation, energy storage, and organ protection. Fats are concentrated in the adipose tissues, which provide cushioning and insulation for parts of the body.

Fats are often grouped into two categories: saturated fats and unsaturated fats. Saturated fats are fats in which the fatty acid backbones only have single covalent bonds, and hence they can be packed tightly and therefore tend to remain solid at room temperature. Animal fats such as butter and lard are examples of saturated fats. Unsaturated fats, to the contrary, are the fats in which the fatty acid backbones have one or more double covalent bonds. These double bonds behave like kinks that prevent the fats from packing tightly, and therefore these fats such as vegetable oils tend to remain in liquid state at room temperature.

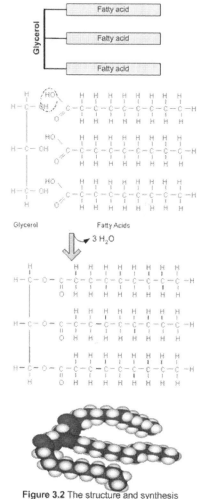

Figure 3.2 The structure and synthesis of a fat molecule called triacylglycerol through dehydration

Omega-3 and omega-6 fatty acids (fats), which are essential for your body and are naturally found in some nuts and vegetables, are examples of unsaturated fats. Omega-3 means the third carbon-carbon bond from the end of the hydrocarbon chain is the double bond.

> Note. Fats with saturated fatty acids are called saturated fats, and fats with unsaturated fatty acids are called unsaturated fats.

The double bonds with kinky behavior are called *cis* bonds, and the unsaturated fats for this reason are also called *cis* fats. Fats with non-kinky (straight) double bonds are called trans fats, and the straight double bond is called a trans bond. Cis and trans forms of an unsaturated fat (or fatty acid) are geometric isomers, which were discussed in Chapter 2. Although some trans fats occur naturally such as in beef, most trans fats are formed in manufacturing processes such as hydrogenation to solidify unsaturated fats. Hydrogenation is usually used to synthetically convert unsaturated fats to saturated (solid) fats by adding hydrogen, but it does produce trans double bonds and hence trans fats. Peanut butter and margarines (hydrogenated vegetable oils) are examples of such trans fats.

> Caution! Trans fats may contribute even more than saturated fats to the cardiovascular disease called atherosclerosis and thereby increase the risk of heart attack.

Phospholipids. An example of complex lipids, phospholipids are lipids with a polar and hence hydrophilic head with a phosphate in it and two nonpolar and hence hydrophobic fatty acid tails. Phospholipids played an essential role in the origin of life and without them there would be no life on Earth as we know it because they synthesize cell membranes. Their dual (hydrophobic and hydrophilic) nature plays a crucial role in membrane synthesis and hence offers an excellent example of *form fits function*.

Steroids. Steroids are lipids with a backbone of four interconnected carbon rings and no fatty acid tails. Eukaryotic cell membranes contain steroids as one of their components. Different steroids are distinguished from one another by the chemical group attached to the fourth ring. Cholesterol is an example of a steroid. Vertebrates obtain it in their diet and synthesize it in their livers. It is a simple steroid from which other steroids such as bile salts, vitamin D, and hormones are formed. Hormones facilitate long-distance signaling in the body of an organism as they travel in body fluid after they are secreted by specialized cells and act on specific target cells.

3.4 Proteins

A protein is one or more polymers of amino acids called polypeptides, which are folded into a specific three-dimensional shape. Amino acids are composed of nitrogen atoms in addition to carbon, hydrogen, and oxygen atoms. Proteins are responsible for almost all dynamic functions of living organisms. Amino acids in a protein come from a set of 20 amino acids called alpha amino acids. Here are some facts about amino acids:

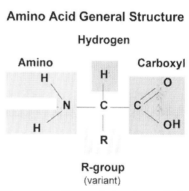

Figure 3.3 General structure of all amino acids

1. As illustrated in Figure 3.3, an amino acid is composed of three components:

Molecular and Cellular Biology

- Carboxyl group
- Amino group
- Side chain called R group that could be nonpolar, polar, or charged. This is what distinguishes one amino acid from another.

2. Polymers (polypeptides) from amino acid monomers are composed through a process called dehydration (Figure 3.4):

   a) The hydroxyl group of amino acid 1 reacts with the hydrogen atom of amino acid 2 to form a molecule of water.

   b) A covalent bond forms between the carbon atom of amino acid 1 and the nitrogen atom of amino acid 2. This bond is called a peptide bond.

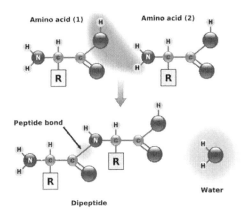

Figure 3.4 Dehydration combining two amino acids during the formation of a polypeptide chain.

   c) Another amino acid can be added to this dipeptide by repeating the two steps above.

   d) Repeated over and over by adding one amino acid at a time, this process gives rise to a chain of amino acids called polypeptides.

A polypeptide, the unique sequence of amino acids, makes the primary structure of a protein.

Molecules of Life

3. The polypeptide chain goes through a process of forming secondary, tertiary, and possibly quaternary structures to form a protein with a specific shape.

An example of a finished product of this process is presented in Figure 3.5, which represents the structure of *myoglobin*, the first protein to have its structure resolved. A technique called X-ray crystallography was used to accomplish this.

The multiple levels of protein structure are described in Table 3.1.

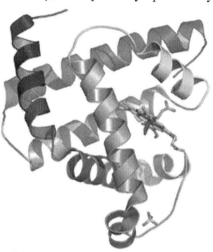

Figure 3.5 A representation of the structure of *myoglobin*, the first protein to have its structure resolved.

Table 3.1 Multiple levels of protein structure.

Structure	Participating Parts and Bonds	Resulting Shapes	Description
Primary	Covalent bonds called peptide bonds between two adjacent amino acids	Linear polymer	The linear polymer of amino acids is also called a polypeptide. Each type of protein has a unique sequence of amino acids.
Secondary	Hydrogen bonds between the repeating constituents of the poly-	α helix and β pleated sheets	α helix is a coil held together by hydrogen bonds between amino acids in the

Molecular and Cellular Biology

	peptide backbone (not the side chains) Interaction between backbone constituents		coil regions next to each other. A β pleated sheet comes into existence as a result of hydrogen bonds between two or more regions (backbones) of the polypeptide chain lying next to (parallel to) each other.
Tertiary	Interaction between side chains (R groups) of the amino acids in the polypeptide backbone Interaction types: hydrophobic and van der Waals interactions between nonpolar side chains, hydrogen bonds between polar side chains, and ionic bonds between positively and negatively charged side chains.	A specific overall three-dimensional shape	Covalent bonds called disulfide bridges also contribute to the tertiary structure.
Quaternary	Weak bonding between two or more polypeptide chains	Overall three-dimensional structure	Only relevant for proteins composed of more than one polypeptide chain.

| | | resulting from the aggregation of multiple polypeptide chains | |

Proteins are composed from a set of 20 amino acids called α-amino acids. As shown in Figure 3.6, these amino acids fall into three categories:

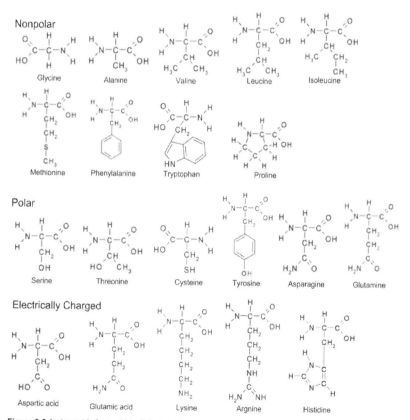

Figure 3.6 Amino acids from which all the proteins necessary to develop and maintain life are composed.

Molecular and Cellular Biology

Nonpolar. These are amino acids with a nonpolar R-group.

Polar. These are amino acids with a polar R-group.

Electrically charged. These are amino acids with an electrically charged R-group.

As Figure 3.6 clearly shows, all amino acids have the same general structure illustrated in Figure 3.4. They can be distinguished from one another by their different R-groups.

Proteins play a variety of roles including biochemical (enzymes), cell signaling (hormones), mechanical (muscle), structural (cytoskeleton), and transport functions. Enzymes are special kinds of proteins that catalyze biological reactions. Here are some other examples of proteins categorized by their functionality:

Contractile and motor proteins. These proteins facilitate cell movement. Examples are *actin* and *myosin*, both of which are responsible for muscle contraction; and *tubulins*, which are components of the spindle fiber that moves chromosomes during cell division.

Defensive proteins. These proteins defend us against disease introduced by foreign entities. Examples of defensive proteins are antibodies, which are produced by the immune system and inactivate or destroy foreign entities such as bacteria and viruses.

Hormonal proteins. We introduced hormones as lipids; however, they can also be proteins called hormonal proteins. Just like lipid hormones, hormonal proteins carry signals in the body to coordinate the activities of an organism. For example, insulin is a hormonal protein secreted by the pancreas which causes liver, muscle, and fat tissues (adipose) to take up glucose from the blood and thereby plays an important role in regulating metabolism.

Receptor proteins. These are proteins that receive chemical signals for a cell coming from outside the cell, and they most often reside on the surface of the cell. Receptors built into the membrane of a nerve cell are examples of receptor proteins.

Storage proteins. These proteins act as storage amino acids. They provide the source of amino acids and can be found, for example, in milk and seeds. *Ovalbumin* is an example of a storage protein; it stores amino acids in egg whites.

Structural proteins. These proteins provide support. Some examples are *collagen* in bones and skin; *fibrin*, which are the fibers of a scab; and *keratin*, which is a component of feathers, hair, horns, and nails.

As shown in Figure 3.7, collagen is a fibrous triple helix protein composed of three coil-like polypeptides wound around each other by hydrogen bonds. It forms cartilage, connective tissues, and the structure of bones, and it accounts for about 40% of all proteins in a human body.

Figure 3.7 A representation of the structure of collagen

Transport proteins. These proteins facilitate the transport of substances in the body of an organism. Some examples are *cytochrome*, which moves electrons through the electron transport system of the cell in crucial cellular processes such as photosynthesis and respiration; *hemoglobin*, which carries oxygen in blood; and *lipoprotein*, which carries cholesterol to the bloodstream.

The *form fits function* principle applies very strictly to proteins, as their functionality depends highly on their shape, which is determined by all levels of their structure. For example, the physical and chemical environment of a protein such as the wrong values of pH and temperature may cause changes in the protein's native shape. This process called *denaturation* affects the functionality of the protein. The wrong shape may also result from a wrong substitution of an amino acid in a polypeptide chain. For example, a blood disorder called *sickle-cell* disease is caused by the substitution of just one wrong amino acid, valine, for the right amino acid, glutamic, at just one position of the polypeptide chain of hemoglobin.

The shape of a protein, which determines the functionality of the protein, is basically determined by its primary structure, the unique sequence of amino acids. What, then, determines the primary structure of a protein? It is coded in genes, which we inherit from our parents. The genes are housed in molecules called DNA, which belong to a class of molecules called nucleic acids. We will explore them next.

3.5 Nucleic Acids

Nucleic acids are large polymers composed of monomers called nucleotides. Two types of nucleic acids, DNA and RNA, are crucial for life. A nucleotide, the monomer of a nucleic acid, is composed of three components:

Nitrogenous base. This is a nitrogen-containing molecule, which due to its chemical characteristic is called a nitrogenous base or just a base. A nucleotide is identified by its nitrogenous base. There are five types of bases relevant to DNA and RNA: adenine (A), cytosine (C), guanine (G), thymine (T), and uracil (U). As illustrated in Figure 3.9, bases C, T, and U have one hexagonal ring of carbon and nitrogen atoms, and they are collectively called pyrimidines. The bases A an G have a double ring, a pentagonal ring fused with a hexagonal ring, and are collectively called purines.

Phosphate group. The phosphate group plays a role in linking one nucleotide to another in a nucleic acid polymer.

Sugar. This is a carbohydrate molecule. Examples of sugars involved in nucleic acids are ribose and deoxyribose; both

Molecules of Life

are pentagonal rings. Sugars also play a role in linking two nucleotides together in nucleic acid molecules.

A nucleotide without its phosphate group is called a *nucleoside*. In a nucleotide, a phosphate group is attached to the sugar, and it links the nucleotide to another nucleotide by making a bond with the sugar of the other nucleotide. This covalent bond is called *phosphodiester* linkage. These linkages compose a chain of nucleotides called *polynucleotide* strand, and sugar-phosphate backbone.

Nucleic acids are polymers of nucleotides that carry the instructions to synthesize proteins, and through proteins for almost all activities of the cell. There are two types of nucleic acids:

Deoxyribonucleic acid (DNA). This is the genetic material that an organism inherits from its parents. DNA is the nucleic acid composed from nucleotides which contains deoxyribose sugar and bases from four types of nucleotides: A, T, C, and G. As illustrated in Figure 3.8, the DNA molecule is a double helix of two phosphate-sugar strands (backbones) curled around each other and connected to each other by hydrogen bonds between their bases. The base A of one strand always bonds with base T of the other strand, and base

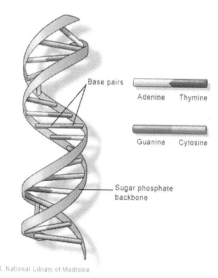

Figure 3.8 Illustration of the double helix structure of a DNA molecule.

Molecular and Cellular Biology

G always bonds with base C. Any two bases bonded together like this in a DNA molecule is called a *base pair*. DNA stores hereditary information that is used to synthesize proteins in a cell. It also has the information and the capability to reproduce (replicate) itself. The double helix structure of DNA facilitates the transfer of information in cell reproduction during cell division.

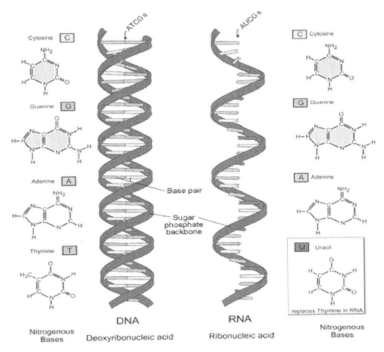

Figure 3.9 Comparison of the structures of DNA and RNA.

Ribonucleic acid (RNA). These molecules are produced according to the instructions in DNA. This nucleic acid is composed of nucleotides which contain ribose sugar and bases from the four types: A, U, C, and G. The RNA molecule usually is a single helix, but two helices can make base pairs: base A with base U and base C with base G. RNA plays a crucial role in protein synthesis in a cell and in gene regulation, and it functions as the genome of some viruses. The comparison of DNA and RNA structures is illustrated in Figure 3.9.

DNA and RNA together enable a generation of a species to hand down the ability to regenerate their complex components to the next generation. You will learn a lot more about DNA and RNA further on in this book.

3.6 In a Nutshell

Molecules of Life in a Nutshell:

- In general, although all the four types of molecules of life perform various (and sometimes overlapping) functions, carbohydrates are energy sources, lipids are structural molecules, proteins are work horses, and nucleic acids are information molecules.

- Out of the four types of molecules of life, carbohydrates, nucleic acids, and proteins are polymers called macromolecules.

- A glycosidic link is a covalent bond between two adjacent sugar monomers in a polymer of a carbohydrate.

- Saturated fats have only single covalent bonds in their fatty acid chains, whereas unsaturated fats include some double bonds.

- An unsaturated fat may come in two isomers: cis with kinky double bonds, and trans with non-kinky (straight) double bonds.

- A polymer of biomolecules such as carbohydrates and proteins forms through the dehydration process in which a water molecule is produced and released, whereas the polymer breaks through the hydration process in which a water molecule is consumed.

Molecular and Cellular Biology

- A peptide bond is a covalent bond between two adjacent amino acids in a sequence of amino acids. These bonds form a polymer from amino acids as monomers, and this polymer is called a polypeptide.
- Polypeptides and proteins are not synonymous. A protein is a polymer that has assumed a specific structure: primary, secondary, tertiary, and possibly quaternary.
- Each type of protein has a unique sequence of amino acids taken from a set of 20 amino acids called α-amino acids.
- Collagen, a fibrous protein composed of three polypeptides, accounts for 40% of the protein in the human body.
- Denaturation is the process in which a protein loses its native unique shape and thereby loses its ability to function properly.
- The monomers in nucleic acid polymers are nucleotides, which are composed of a phosphate group, sugar, and nitrogenous bases.
- DNA is almost always a double helix of polynucleotide chains, whereas RNA is usually a single helix.
- The bond that links two nucleotides together in a nucleic acid are covalent bonds called phosphodiester linkages, whereas the bonds that link two polynucleotides into a double helix are hydrogen bonds made between the bases of the two polynucleotide strands.
- DNA carries hereditary information.

Some important facts about the molecules of life are summarized in Table 3.2.

Table 3.2 Molecules of life: some facts.

Type of Molecules	Components/Monomers	Non-methyl functional groups in the monomer	Covalent bonds between monomers
Carbohydrates	Monosaccharides; also called simple sugars	Hydroxyl, carbonyl	Glycosidic linkage
Lipids	Glycerol and fatty acids for fats, backbone of four fused carbon rings for steroids	Hydroxyl	Ester linkage (in fats)
Proteins	Amino acids	Carboxyl, amino	Peptide bond
Nucleic acids	Nucleotides	Phosphate	Phosphodiester linkage

3.7 Review Questions

1. Which of the following biomolecules contain nitrogen in addition to carbon, hydrogen, and oxygen?

    A. fructose

    B. chitin

    C. insulin

    D. cholesterol

    E. DNA

Molecular and Cellular Biology

F. all of the above

G. B., C., and E.

2. A bond between two monomers in a complex carbohydrate polymer is called a ____.

    A. peptide

    B. glycosidic link

    C. phosphodiester bond

    D. covalent bond

    E. A. and D.

3. Humans and other vertebrates store their energy in their ____ in the form of ____.

    A. livers, glycogen

    B. livers, cellulose

    C. plastids, starch

    D. livers, starch

4. The exoskeleton of a spider is largely made of ____.

    A. collagen

    B. chitin

    C. glycogen

    D. cellulose

5. Which of the following is not a true statement about trans fats?

    A. They may contribute even more than fatty acids to heart attack.

B. They are unsaturated fats.

C. They have no double covalent bonds.

D. They are isomers of cis fats.

6. What is true about unsaturated fats?

    A. They have only double covalent bonds in their fatty acid chain.

    B. They have only single covalent bonds in their fatty acid chains.

    C. They may come in two forms: cis and trans.

    D. They can never be converted to saturated fats.

7. Which of the following is the process in which a protein loses its native shape and thereby loses its ability to function properly?

    A. dehydration

    B. hydrogenation

    C. overheating

    D. denaturation

8. A bond between two monomers in a polynucleotide strand of a nucleic acid polymer is called ____.

    A. peptide

    B. glycosidic link

    C. phosphodiester bond

    D. hydrogen bond

9. Two polynucleotide strands link with each other to make a double helix through which bond?

Molecular and Cellular Biology

   A. glycosidic link
   B. phosphodiester bond
   C. hydrogen bond
   D. peptide bond

10. What makes an amino acid, an acid?

   A. carboxyl group
   B. amino group
   C. R-group
   D. phosphate group

3.8 Answer Key

1. G.
2. E.
3. A.
4. B.
5. C.

6. C.
7. D.
8. C.
9. C.
10. A.

Notes:

Q6. Unsaturated fats can be synthetically converted to saturated fats by a manufacturing process called hydrogenation, which adds hydrogen to the fatty acid chains.

Chapter 4

Anatomy and Physiology of Cells

4.1 Biological Cells: The Big Picture

Cells (Figure 4.1) are anatomically and physiologically the fundamental units of all living organisms. The word *cell* originated from the Latin word *cellula*, which means a small room. Cell theory, supported by microscopy and other evidence, consists of the following three generalizations:

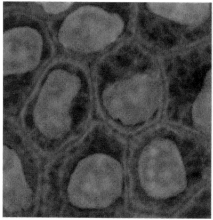

Figure 4.1 Cells in culture stained for protein in the cell membrane (red) and DNA in the nucleus (green).

1. All organisms consist of one or more cells.

2. The cell is the smallest unit that retains the properties of life.

3. Each new cell arises from an already existing cell.

Molecular and Cellular Biology, by Paul Sanghera
Copyright © 2015 Infonential.

Molecular and Cellular Biology

As you know from Chapter 1, all life on Earth can be grouped into three domains: Archaea, Bacteria, and Eukarya. Similarly, all cells can be broadly grouped into two categories: prokaryotic cells, which compose all organisms in Domains Archaea and Bacteria called prokaryotes, and eukaryotic cells that compose all organisms in the Domain Eukarya called eukaryotes. A eukaryotic cell is the cell type that has membrane-bound internal structures called organelles and a membrane-bound nucleus in which DNA is stored. In contrast, a prokaryotic cell is the cell type that lacks these membrane-bound internal structures and a membrane-bound nucleus.

The prokaryotic cell is much simpler in structure than the eukaryotic cell because it lacks the nucleus and organelles. Typical bacteria (prokaryotic cells) are 1-5 μm in diameter, whereas typical eukaryotic cells are 10-100 μm in diameter. The smallest known type of bacteria are mycoplasma, which vary in size from 0.1 μm to 1 μm.

> **Note.** You can see cells through a light microscope, but you cannot see the details of the internal structure of a cell by using the light microscope because its resolution is limited by the wavelength (or frequency) of visible light. However, you can see the details of the internal structure of a cell by using an electron microscope because electron microscopes offer better resolution to the level of a few nanometers due to the smaller wavelength of electron beams than that of visible light.

While we further explore the differences between different types of cells, do not lose sight of the fact that all cells have a lot of things in common, including some very fundamental things. For example, all cells perform most of the same kinds of functions. All cells have an outer boundary called a plasma membrane, which encloses a semi-fluid matrix called cytosol loaded with internal structures called ribosomes. Not only do all cells have DNA, but also the DNA they have

Anatomy and Physiology of Cells

carries identical genetic code. These similarities stem from the fact that all life on Earth evolved from a common ancestor.

> Caution! Although the basic chemistry of cells can be understood in terms of the structures and functions of molecules of life (carbohydrates, lipids, proteins, and nucleic acids), cells also have other substances in them such as water molecules, which account for 70% of the total cell mass, and inorganic ions such as sodium, potassium, and calcium ions, which make about 1% of the cell mass.

At this point, you can ask a question: What really drives the size and shape of a cell? For example, why are cells so small? The lower limit on size (how small a cell can be) is driven by the requirement that all necessary metabolic components must fit within the cell. The upper limit (how large a cell can be) on size is driven by the need for efficient exchange of metabolic materials with the environment, which in turn gives rise to the requirement of high enough *surface area to volume* ratio. The surface must be large enough to serve the enclosed volume in transferring material to and from the volume. The requirement of surface area to volume ratio also affects the shape of a cell.

In this chapter, we will describe the anatomy and physiology of biological cells by exploring and comparing both types of cells: prokaryotic and eukaryotic. We will also briefly explain the difference between animal and plant cells.

4.2 Anatomy and Physiology of Prokaryotic Cells

The main components of a typical prokaryotic cell structure are shown in Figure 4.2 and are described briefly in the following.

Cell wall. This is a rigid yet permeable structure surrounding the plasma membrane in almost all prokaryotic cells. It helps a cell resist rupturing and retain its shape.

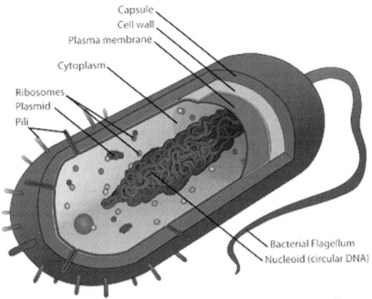

Figure 4.2 Illustration of a typical prokaryotic cell.

Capsule. This is a slim layer of sticky polysaccharides around the cell wall. It helps unicellular organisms such as bacteria to adhere to the surface and to protect themselves from predators and toxins.

Cytoplasm. This is the content of the cell in between the plasma membrane and its nucleoid (in a prokaryotic cell) or nucleus (in a eukaryotic cell). The fluid (or semi-fluid) part of the cytoplasm is called cytosol.

Flagellum. This is a long slender cellular structure used by bacteria for moving.

Nucleoid. This is a region in the prokaryotic cell that contains the chromosomal DNA of the organism. This region is not membrane bound.

Plasma membrane. Also called cell membrane, this is the outer membrane of all cells that encloses the cytoplasm of the cell. It acts as a selective barrier to determine what can go into and out of the cell.

Plasmid. This is a small circular DNA molecule inside the cell that carries genes separate from the bacterial chromosome. It is not essential for cell survival, but it is still useful, for example, in DNA cloning. Plasmids are also found in some eukaryotic cells such as in yeasts.

Ribosomes. These are the sites in the cells where proteins are synthesized. A ribosome is composed of a type of RNA called ribosomal RNA (rRNA) and proteins.

Pili. These are protein filaments projecting from the surface of the cell. They help bacteria cling to or move along a surface. These kind of pilis are called *fimbriae*. The other type of pili are called sex pili. A sex pilus is so named because a bacterium uses it to link to another bacterium in order to transfer genetic material. Fimbriae are shorter than sex pili.

Sex pilus, as mentioned earlier, is a kind of pili that is used by a bacterium to transfer plasmid (genetic material) directly into another bacterium. The donor cell attaches its sex pilus to the recipient cell, shrinks it to draw the recipient closer, and injects a copy of its plasmid into the recipient. This process is called *conjugation*.

Prokaryotic cells were the first cells that appeared on Earth, and eukaryotic cells evolved from prokaryotic cells.

4.3 Anatomy and Physiology of Eukaryotic Cells

In this section, we will explore the anatomy and physiology of a typical animal cell as an example of a eukaryotic cell. Suspended in the cytoplasm of a eukaryotic cell are membrane-bound structures with specialized functions. These structures are commonly called organelles. As already mentioned, prokaryotic cells lack organelles. Organelles are very crucial evolutionary traits that distinguish eukaryotic cells from prokaryotic cells.

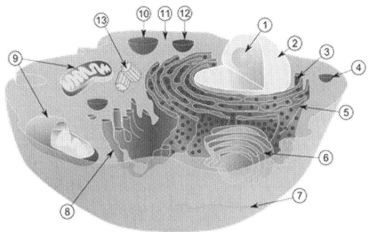

Figure 4.3 Illustration of a typical eukaryotic cell.

Figure 4.3 illustrates a typical animal (eukaryotic) cell, showing its internal structures, mostly the organelles. In the following, we describe these structures by the same number as assigned to them in the figure.

1. **Nucleolus.** This is a round structure within the nucleus that contains the genes to produce rRNA, which is syn-

Anatomy and Physiology of Cells

thesized in this structure. Ribosomal units are synthesized in this structure by using the rRNA synthesized here and some proteins imported from the cytoplasm of the cell.

2. Nucleus. This is an organelle which contains genetic material, DNA, packed in chromosomes. It has an outer envelope that separates chromosomes inside the nucleus from the cytoplasm.

3. Ribosomes. These are sites where proteins are synthesized. In a eukaryotic cell, they are assembled in the nucleus and exported to the cytoplasm.

4. Vesicle. This is a saclike membrane-bound structure in the cytoplasm of the cell. Vesicles in a cell store and transport cellular substances and can also modify or degrade these substances.

5. Rough endoplasmic reticulum (ER). Endoplasmic reticulum (ER) is a continuous membranous network of saclike and tubelike structures in eukaryotic cells, and it exists as an extension of the nuclear envelope. Smooth ER is the region of ER free of ribosomes, and rough ER is the region of ER that is studded with many ribosomes. Transport vesicles bud off from a region of rough ER and stretch to the Golgi apparatus in order to transport material.

6. Golgi apparatus. This is composed of stacks of flattened sacks of membranes. This is an organelle in which the products made in the cell are sorted and packaged for transportation. Some products are also modified here such as some proteins and phospholipids, and some products are even synthesized here such as many polysaccharides.

Molecular and Cellular Biology

7. **Cytoskeleton.** This is a dynamic framework of diverse protein filaments and microtubules that extends throughout the cytoplasm. It performs a wide spectrum of mechanical, signaling, and transport functions. For example, it plays a significant role in organizing, structurally supporting, and moving the cell and its internal structures.

8. **Smooth endoplasmic reticulum (ER).** This is the region of ER that is not studded with ribosomes. Its functions include synthesis of lipids, metabolism of carbohydrates, detoxification of drugs and other poisonous substances, and storage of calcium ions (Ca^{2+}).

9. **Mitochondrion.** This is a double membrane organelle. It functions as the site of cellular respiration in which oxygen is consumed to produce cellular energy in the form of ATP (adenosine triphosphate) molecules by breaking down organic molecules. In eukaryotes, it is also the site of the second and third stages of aerobic respiration.

10. **Vacuole.** This is a membrane-bound organelle, which is basically a large vesicle. It performs varied functions in different types of eukaryotic cells including cell growth, digestion, maintaining pH, protection, storing molecules, waste disposal, and maintaining the shape and structure of the cell.

11. **Cytoplasm.** This is the semi-fluid content of the cell, which includes water, sugar, proteins, and ions, between a cell's plasma membrane and its nucleus (in a eukaryotic cell) or nucleoid (in a prokaryotic cell).

12. **Lysosome.** This is a vesicle (in animal cells) that buds from the Golgi body (apparatus) and is filled with enzymes that facilitate intracellular digestion. Because these enzymes are capable of breaking down all the molecules of life, their

compartmentalization into lysosomes prevents the general destruction of cellular components by them. There are similar organelles called peroxisomes which remove hydrogen atoms from various substances and combine them with oxygen atoms to make hydrogen peroxide (H_2O_2) and subsequently turn this toxic substance into water by using enzymes that they contain. Peroxisomes use this functionality to perform various tasks such as breaking fatty acids to smaller molecules, which by going through mitochondria are eventually used as a fuel in cellular respiration, and they detoxify harmful substances such as alcohol in the liver.

13. **Centrioles within centrosome.** A centrosome is a region in the cytoplasm in which microtubules form, whereas a centriole is a barrel-shaped structure that organizes newly forming microtubules. Microtubules are hollow rod-like nanostructures with diameter about 25 nm and length varying from 200 nm to 25 μm. These tubes shape and support the cell and also provide the routes for some organelles to move around in the cell.

> **Fascinating Fact!** Mitochondria and chloroplasts have their own ribosomes and circular DNA molecules. They use these DNA and ribosomes to synthesize some of the proteins that they need. This is evidence that eukaryotes evolved from prokaryotes by engulfing other prokaryotes.

Here are some prominent differences between plant cells and animal cells, both eukaryotes:

Molecular and Cellular Biology

What animal cells have and plant cells do not:	What plant cells have and animal cells do not:
• Flagella (present in some plant sperms) • Centrosomes with centrioles • Lysosomes	• Cell wall • Chloroplasts • Central vacuole • Plasmodesmata

Plasmodesmata are open channels through plant cell walls that connect the cytoplasm of adjacent cells. This opens the communication channels between adjacent cells as the molecules can pass through the plasmodesmata from one cell to another. The equivalent of plasmodesmata in animal cells is called *gap junctions*.

The ER, Golgi apparatus, lysosomes, nuclear envelope, and vacuoles discussed in this section are part of a system called the endomembrane system, which is vital to the functioning of the cell. Let us explore it further.

Note. Photosynthetic eukaryotic cells contain a family of closely related organelles called plastids, which include chloroplasts, chromoplasts, and amyloplasts.

4.4 The Endomembrane System

The endomembrane system (Figure 4.4) is the system comprised of all the membranes inside and surrounding a eukaryotic cell. These membranes make a continuous system; that is, they are interrelated either through physical links or through moving membranous vesicles. This system includes the nuclear envelope, ER, Golgi apparatus, lysosomes, vacuoles, and plasma membrane. The endomembrane system performs a set of significant tasks including detoxification,

Anatomy and Physiology of Cells

formation and movement of lipids, metabolism of carbohydrates, and synthesis of proteins and protein transportation. These proteins may become part of the membranes, travel to organelles, or be secreted out of the cell.

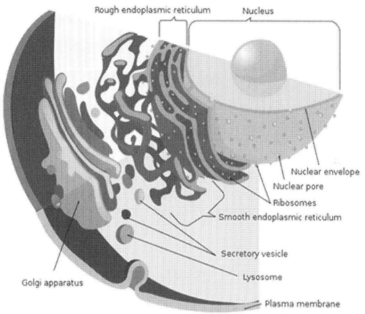

Figure 4.4 Illustration of a typical endomembrane system

In the following, we describe the process of synthesis and transportation performed by the endomembrane system, as illustrated in Figure 4.5.

1. It begins in the cell nucleus. The genetic instructions are carried by DNA molecules organized into discrete units called chromosomes. These instructions are transcribed into structures called messenger RNA (mRNA), which carry these instructions to ribosomes. The structural units of ribosomes themselves are synthesized inside the nucleus from rRNA and proteins, which then exit the nucleus into the cytoplasm through nuclear pores and assemble themselves into

Molecular and Cellular Biology

ribosomes. The mRNA, carrying the DNA instructions to synthesize protein, exits the nucleus through nuclear pores.

2. The Next Action at Ribosomes. As previously mentioned, ribosomes are the complexes that carry out protein synthesis. The genetic instructions carried by the mRNAs to ribosomes are translated into proteins here. Ribosomes can freely float inside the cytosol, or they may be bound to the outside of a rough ER or nuclear envelope. Some ribosomes can switch between two states: free and bound. Ribosomes in cytosol, the free ribosomes, make most of the proteins used in the cytosol. Proteins that will become part of the membranes, will be packaged for organelles, or will be secreted out of the cell are synthesized by the ribosomes bound to the outside of rough ER.

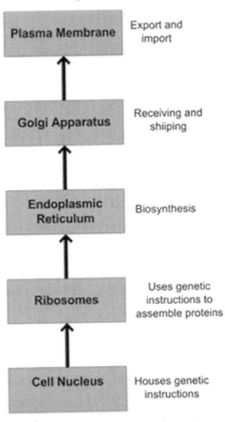

Figure 4.5 Functioning of a cell

3. Processing Through Endoplasmic reticulum (ER). Recall that the ER is a network of tubelike and saclike membranous structures that can carry out biosynthesis.

a) Rough ER modifies proteins. These proteins will either become part of the membranes of the organelles in the cell or will be secreted out of the cell.

b) Proteins that reach smooth ER specialize into enzymes or membrane proteins. These enzymes play an important role in synthesizing lipids such as oil, phospholipids, and steroids. Sex hormones produced in animal cells are an example of steroids. In addition, some of these enzymes participate in diverse metabolic processes such as metabolizing carbohydrates and detoxifying drugs and poisons.

Some proteins after leaving the ER travel to the Golgi apparatus through the vesicles.

4. Golgi apparatus, the Shipping Center. Products of the cell such as proteins that arrive here from the ER are modified, stored, packaged, and sent to other destinations. In other words, the Golgi apparatus receives, processes, repackages, and ships products that arrive here from the ER.

5. Plasma membrane, the Gateway to and from the Cell. Some vesicles called exocytic vesicles bud from the Golgi membrane and fuse with the plasma membrane, the outer boundary of the cell. This way, they transport cell products and wastes to the plasma membrane, which in turn releases them to the outside of the cell.

Some other vesicles called endocytic vesicles form a patch of the plasma membrane and sink into the cytoplasm of the cell. They move some useful substances such as nutrients and water into the cytoplasm of the cell from the outside of the cell.

Molecular and Cellular Biology

4.5 Summary: Putting It All Together

The cell, the fundamental structural and functional unit of life, is made of and makes molecules of life, which in turn are composed of atoms of chemical elements. The cell, however, is not just a system of components; it is a self-sustained chemical factory continuously run by its own components summarized in Table 4.1.

Table 4.1 Components of a cell.

Cell Component	Major Structural Elements	Function
Cell Wall Protective layer external to plasma membrane in the cells of plants, prokaryotes, fungi, and some protists	Polysaccharides (such as cellulose in plants and some protists and chitin in fungi), carbohydrates, and amino acids (such as peptidogly can in bacteria) .	Protects the cell, maintains its shape, and prevents excessive uptake of water
Plasma Membrane The membrane at the boundary of every cell enclosing the cell	Lipids and proteins	Acts as a selective barrier to regulate the cell's chemical composition
Cytoskeleton A network of fibers extending throughout the cytoplasm	Microtubules, cilia, and flagella	Gives mechanical support to the cell and maintain its shape
Mitochondria Membrane enclosed organelle that works as energy factory for	Proteins including enzymes, ribosomes, and DNA	Cellular respiration Converts chemical energy from food (organic compounds)

the cell		to the form that the cell can use (ATP)
Chloroplast In animal cells, not plant cells	DNA, ribosomes, chlorophyll, and proteins (enzymes)	Runs photosynthesis to convert CO_2 and water into organic compounds such as carbohydrates by using sunlight
Endoplasmic Reticulum (ER) Network of membrane-bound tubules and sacs	Membranous sacs and tubes, lipids, and proteins	Rough ER: synthesis of proteins by ribosomes bound to the ER. Smooth ER: Synthesis of lipids, metabolism of carbohydrates, detoxification of drugs, and Ca^{2+} storage
Golgi Apparatus Sorting and shipping of cell products including proteins	Membranous sacs	Synthesizes, modifies, and sorts cell products for secretion
Lysosome Only in animal cells	A membranous sac of hydrolytic enzymes	Breaks down macromolecules of the cell, ingested material, and damaged organelles
Vacuole Only in plants	Membrane-bound vesicle	Storage of molecules and ions such as proteins, potassium, and chloride Also plays a part in digestion and contributes to cell growth

Molecular and Cellular Biology

Peroxisome Metabolic compartment	Enzymes	Performs some metabolic functions such as the transfer of hydrogen to oxygen, which produces hydrogen peroxide, a toxic substance. Converts hydrogen peroxide to water
Ribosome Free ribosomes in cytosol and ribosomes bound to the rough ER	rRNA and proteins	Synthesizes protein
Nucleus Defining component of eukaryotic cells.	Chromatin, DNA, and proteins	Houses chromosomes which contain DNA and synthesizes ribosomal subunits

Note that both mitochondria and chloroplasts function as energy transformation centers. Mitochondria transforms the energy of organic compounds in food into the form of ATM molecules that the cell can use, whereas chloroplasts transfer sunlight energy into the form of organic compounds such as carbohydrates synthesized from carbon dioxide and water through a process called photosynthesis.

Table 4.2 presents a comparison of prokaryotic and eukaryotic cells in terms of their typical components.

Table 4.2 Typical components of prokaryotic and eukaryotic cells.

Cell Component	Prokaryotic — Bacteria Archaea	Eukaryotic — Animals	Fungi	Plants	Protists
DNA, ribosomes, and plasma membrane	✓	✓	✓	✓	✓
Cytoskeleton, ER, Golgi apparatus, mitochondrion, nucleus	✗	✓	✓	✓	✓
Cell wall	✓	✗	✓	✓	✓
Chloroplast	✗	✗	✗	✓	✓ some
Central vacuole	✗	✗	✓ some	✓	✗
Lysosome	✗	✓	✗	✗	✓ some

Molecular and Cellular Biology

4.6 In a Nutshell

Similarities and differences among different cells present another piece of evidence that supports the theory of evolution, which is descent with modification from a common ancestor. Here is a brief summary of some facts about biological cells.

- New cells are created from existing cells.

- All cells have a plasma membrane, the outer boundary of the cell used for selective export and import; ribosomes are used to synthesize proteins; and DNA contains hereditary information. All cells also have cytosol.

- Prokaryotic cells do not have a nucleus or any membrane-bound organelles, but they do have a nucleoid and ribosomes.

- Prokaryotic cells do not have internal membranes. Any membrane being used inside a prokaryote cell is an extension of the plasma membrane.

- Here is a way to distinguish between cytosol and cytoplasm: Cytosol is the semifluid jellylike substance enclosed by the plasma membrane, whereas cytoplasm is the region between the nucleus and the plasma membrane. Thus, cytoplasm contains organelles and other cell components suspended in cytosol.

- Animal cells, not plant cells, have centrosomes (regions where microtubules are initiated) and a lysosome, a digestive organelle.

- Plant cells, not animal cells, have chloroplasts, a central vacuole, a cell wall, and plasmodesmata. The central vacuole performs the function in a plant cell that a lysosome performs in an animal cell.

- Even in a eukaryotic cell, while most of the DNA is in the nucleus, mitochondria and chloroplasts have their own DNA and ribosomes too, which they use to produce some of their own proteins. However, most of the proteins for these organelles are imported from free ribosomes in the cytoplasm.

- The internal structures called ribosomes have no membrane enclosing them, and therefore they are not considered organelles.
- Gap junctions in animals and plasmodesmata in plants provide cytoplasmic channels between adjacent cells to facilitate communication between the cells such as exchanging material.
- The cellular membranes of a cell make an integral continuous system called the endomembrane system, which facilitates metabolic functions and traffic of organic molecules such as proteins.

4.7 Review Questions

1. When we say that the cell is the fundamental building block of life, we mean it has a capability for ____.

   A. growth

   B. homeostasis

   C. metabolism

   D. reproduction

   E. all of the above

   F. A., B., and C.

2. Which of the following statements about a prokaryotic cell is not true?

   A. It has a nucleoid in which the chromosomal DNA resides.

Molecular and Cellular Biology

B. It has ribosomes.

C. It does not have any internal structure.

D. It has a cell wall.

3. The chromosomal DNA in prokaryotic cells is concentrated in a region called the ____.

   A. nucleus

   B. nucleoid

   C. plasmid

   D. nucleolus

4. All of the following are components of a prokaryotic cell except for ____.

   A. cell wall

   B. nucleoid

   C. ribosome

   D. mitochondria

   E. chromosomal DNA

5. Which of the following is a factor in determining the cell size?

   A. sufficient surface area for the cell to function in the enclosed volume

   B. the fact that more complex cells evolved from the simpler cells

   C. the fact that prokaryotes do not have any membrane-bound internal structures

D. the fact that all types of cells need DNA for self-production

6. Which of the following is not true about ribosomes?

   A. They exist in cytosol.

   B. They exist in rough ER.

   C. They exist in smooth ER.

   D. They are the sites for protein synthesis.

7. Ribosomes in cells serve as production sites for what kind of molecules?

   A. DNA

   B. proteins

   C. lipids

   D. carbohydrates

   E. all of the above

8. Which of the following organelles have their own DNA and ribosomes?

   A. Golgi apparatus

   B. lysosome

   C. mitochondrion

   D. chloroplast

   E. C. and D.

9. Which of the following organelles convert energy from one form to another?

Molecular and Cellular Biology

   A. Golgi apparatus
   B. lysosome
   C. mitochondrion
   D. chloroplast
   E. C. and D.

10. The proteins that may be exported from the cell are synthesized in ____.
    A. rough ER
    B. smooth ER
    C. the Golgi apparatus
    D. peroxisomes

11. Toxic substances are neutralized (detoxified) in the liver by ____.
    A. lysosomes
    B. peroxisomes
    C. Golgi vesicles
    D. ribosomes

12. Which of the following statements about cells is false?
    A. All cells have a plasma membrane.
    B. All cells have DNA.
    C. Some biotechnology companies manufacture cells from molecules of life and sell them to scientists in the form of cell cultures.
    D. All cells use the same genetic code.

4.8 Answer Key

1. E.
2. C.
3. B.
4. D.
5. A.
6. C.
7. B.
8. E.
9. E.
10. A.
11. B.
12. C.

Notes:

Q2. A prokaryotic cell may have internal structures but not membrane-bound internal structures called organelles.

Q12. Each new cell arises from an already existing cell.

Molecular and Cellular Biology

Chapter 5

Cell Membrane: The Supermolecule of Life

5.1 Cellular Membranes: The Big Picture

Cellular membranes facilitate life. Molecules of life, which are discussed in Chapter 3, are enclosed by a cellular membrane called plasma membrane or cell membrane. A cellular membrane is a biological material boundary that encloses a solution different from the surrounding solution and selectively controls the traffic into and out of the area it surrounds. The property of the membrane to selectively control the incoming and outgoing traffic is called *selective permeability*.

The plasma membrane, which is illustrated in Figure 5.1, separates each living cell from its surroundings. We will explore its structure including the components labeled in this figure further on in the chapter. In addition to the plasma membrane, a eukaryotic cell also has internal cellular membranes that form the boundaries of its organelles and provide compartmentalized specialization in the cell. Plasma membrane and internal membranes are very similar in their

Molecular and Cellular Biology, by Paul Sanghera
Copyright © 2015 Infonential.

Molecular and Cellular Biology

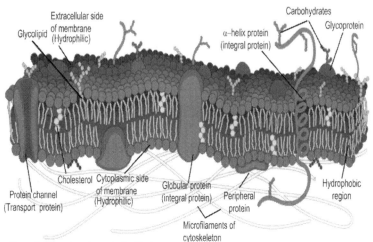

Figure 5.1 Illustration of an animal cell's plasma membrane

composition and structure. As you learned in Chapter 4, all these cellular membranes of a cell form an integral continuous system called the *endomembrane system*, which facilitates metabolic functions and traffic of organic molecules such as proteins.

> **Fascinating Fact!** Plasma membrane is typically only 8 nanometer (nm) thick, and the polar head of a phospholipid in it is about 0.8 nm in length. The typical membrane of mitochondria, an internal membrane, is about 6 nm thick.

The property of selective permeability of membranes provides a cell the ability to be selective in exchanging substances with its external environment and to maintain an internal chemical environment different from its surroundings. This ability has been instrumental in the origin, survival, and evolution of life.

The anatomy and physiology of cellular membranes are some of the countless examples of the *form fits function* principle.

5.2 Anatomy and Physiology of Cellular Membranes

Major anatomical and physiological features of cellular membranes are described in this section.

Structure. According to the fluid mosaic model, supported by evidence, cellular membranes are fluid mosaics of lipids and proteins. They consist of a phospholipid bilayer embedded with proteins. Also according to this model, most of the phospholipids and some of the proteins can move laterally in the membrane. As shown in Figure 5.1, each of the two lipid layers in the membrane is composed of phospholipids. A phospholipid has a hydrophilic (polar) head with a phosphate in it and two hydrophobic fatty-acid tails. Because water is polar, the polar heads of the external layer of phospholipids are attracted to the aqueous environment external to the cell, and the polar heads of the internal layer are attracted to the aqueous environment of cytosol. The hydrophobic tails of one layer are attracted to the hydrophobic tail of the other layer, and the tails in both the layers are repellent to the aqueous environments external and internal to the cell. These forces of attraction and repulsion create the bilipid layer arrangement for membranes, as shown in Figure 5.1.

> Note. In organic chemistry, there is a rule called *like attracts like* or (or hence) *like dissolves like*. This means that polar substances are attracted to other polar or hydrophilic substances and nonpolar substances are attracted to other nonpolar or hydrophobic substances, but polar and nonpolar substances are repellent to each other. For example, urea, a highly polar substance, is very soluble in water, which is also highly polar. However, urea is not soluble in benzene, which is a nonpolar solvent. In contrast, the nonpolar solute naphthalene is highly soluble in benzene, but naphthalene is not soluble in water.

Molecular and Cellular Biology

Synthesis. Where do the lipids and proteins of a cellular membrane come from? They are made in the endoplasmic reticulum (ER), which was discussed in Chapter 4. To be specific, these proteins are made at ribosomes attached to the rough ER and modified in the Golgi apparatus before they arrive at a specific membrane. Lipids including oils, phospholipids, and steroids are made in the smooth ER and modified in the Golgi apparatus. As described earlier, the phospholipids spontaneously make a bilipid layer arrangement.

Fluidity. A cellular membrane is not a rigid structure of molecules. Phospholipids and to some extent proteins too move laterally within the membrane. The phospholipid tails in the interior of a membrane if composed of unsaturated hydrocarbons increase fluidity and help keep the membrane fluid even at low temperatures. Furthermore, by playing a dual role, cholesterol in the membrane resists changes in the fluidity of the membrane even when the external temperature changes. The cholesterol molecules are wedged in between the phospholipid molecules. At a moderate or relatively high temperature, the cholesterol molecules reduce membrane fluidity because they constrain the motion of the phospholipid molecules. On the other hand, as the temperature lowers, the cholesterol molecules begin enhancing fluidity by resisting solidification as they disrupt the packing of phospholipids. Thus, the dual role of cholesterol helps to stabilize phospholipids and hence cellular membranes by functioning as a fluidity buffer resisting the change in fluidity due to a change in temperature.

> Fascinating Fact! Adjacent phospholipid molecules in a cellular membrane switch positions about ten million times every second. Protein molecules, however, move within the membrane much slower because they are larger than phospholipids.

Structural variations. So far we have talked about the global structure of all membranes. However, different membrane-bound structures in the cell differ in their functions, and their membranes vary in their structure to support those functionalities. Even the two sides (external and internal) of a cellular membrane are different to serve different functions. In other words, the exterior and interior faces of the bilipid layer have different molecular compositions. For example, unlike the interior face, the exterior face has some short carbohydrate chains bonded to lipids or proteins, and these chains project into the extracellular environment. These molecules communicate or interact with the surface molecules of other cells, and this communication facilitates *cell-to-cell recognition*.

Selective permeability. Because the interior of a membrane is composed of the nonpolar tails of phospholipids, nonpolar (hydrophobic) entities pass rapidly through the membrane. However, polar molecules and ions need assistance from proteins called transport proteins to pass through the membrane. With its ability to control the passage of molecules and ions, membrane structure naturally leads to the selective permeability characteristic of the membrane.

5.3 Role of Proteins in Cellular Membranes

While phospholipids largely form the fabric of a cellular membrane, proteins define most of its functions. Not only do different types of cells contain different types of membrane proteins, but also various internal membranes of the same cell contain different sets of proteins to support their different functions. From their location in the membrane, proteins can be broadly grouped into two categories:

Molecular and Cellular Biology

Integral proteins. An integral protein is a membrane protein that either spans the whole membrane by penetrating through the hydrophobic core of the membrane or penetrates partway into the hydrophobic core from either side of the membrane. A *transmembrane* protein is an integral protein that usually spans the membrane several times, and its hydrophobic parts have a helical structure as shown in Figure 5.1.

Think About It!

For a protein to be an integral protein, should it be hydrophobic, hydrophilic, or both?

Answer:

An integral protein must have both hydrophobic and hydrophilic parts because it links with the hydrophilic environment external to the cell and passes through the hydrophobic interior of the membrane. An entity that has both hydrophobic and hydrophilic regions is called amphipathic.

Peripheral proteins. A peripheral protein is a membrane protein that is loosely bound to the surface of the membrane and not embedded in it.

These proteins perform diverse functions. The major membrane protein categories based on their functions are described in the following.

Attachment, shape, and coordination. These membrane proteins help maintain cell shape by noncovalently binding to the cytoskeleton and extracellular matrix (ECM). Some of them also coordinate changes inside and outside of the cell by binding to the molecules in the extracellular matrix.

Cell-to-cell recognition. These proteins on the surface of a cell, which are shown as glycoproteins in Figure 5.1, are recognized by the proteins on the membranes of other cells and thereby facilitate cell-to-cell recognition. A glycoprotein is a protein that has one or more carbohydrates attached to it by covalent bonds.

Enzymatic function. These proteins are enzymes. Their active site is exposed to the reactant molecules in a solution such as cytosol so that they can catalyze biological reactions. For example, a set of enzymatic proteins embedded in the membrane facilitates the sequential steps of a metabolic pathway.

Intercellular joining. These proteins on a cell bind to the proteins on an adjacent cell and this way link the two cells. As a result, these membrane proteins can form various junctions between adjacent cells such as a gap junction and a tight junction. A gap junction is a junction that permits the flow of ions and some molecules of small enough sizes between the cytoplasm of the two cells. However, a tight junction is a cell junction that prevents such a flow.

Signal transduction. These proteins receive a message from an external chemical messenger such as a hormone and relay it to the inside of the cell. For example, a protein offers a binding site to which an external chemical messenger such as a hormone binds. This binding causes a shape change in the protein, and this shape change helps the protein to bind to a cytoplasmic protein and relay the message this way. This relaying of message begins a series of steps, called the signal transduction pathway, that will ultimately result in a response to the original extracellular chemical signal.

Transport proteins. These proteins help entities such as molecules and ions transport through the membrane. Examples of channel proteins are shown in Figure 5.1.

Transportation through membranes plays an instrumental role in cell functioning and is therefore discussed further in the next section.

5.4 Transportation Through a Membrane

In the following, we describe various transport-related processes and the working of membrane proteins in the process of transportation through a membrane.

Diffusion. Diffusion is the spontaneous movement of entities such as molecules down the concentration gradient from a region where the entities are more concentrated to the region where they are less concentrated. It is a natural tendency for molecules of any substance to spread out evenly into the available space. This means no energy is expended in diffusion. If there are multiple kinds of entities in a given space, each kind diffuses down its own concentration gradient unaffected by the concentration of any of the other entities.

Passive transport. Passive transport is the diffusion of entities along a biological membrane. In other words, passive transport is the spontaneous movement of entities across a biological membrane along (or down) the electrochemical gradient. No energy is consumed in this process. An electrochemical gradient is the gradient built by the difference in the concentration of the entities such as the number of ions across the membrane and the electric force across the membrane between the external and internal entities.

Facilitated diffusion. Passive transport facilitated by membrane proteins is called facilitated diffusion. Even though it is assisted by membrane proteins, no energy is spent in facilitated diffusion.

Active transport. This is the transportation of an entity across a membrane against the electrochemical gradient of the membrane. Obviously, an expenditure of energy is required to facilitate this process.

Transport protein. This is a transmembrane protein that facilitates the transportation of a specific kind of entity (or very closely related kinds of entities) across a membrane. As shown in Figure 5.2, channel proteins and carrier proteins are two types of transport proteins.

Channel protein. This is any transport protein that provides a corridor (or channel) through which an entity (a specific molecule or ion) can cross a membrane. Channel proteins only facilitate passive transport. An example is *aquaporin*, which is a channel protein that facilitates the flow of water through the membrane. Water can move through the membrane without aquaporins, but it does so relatively slowly.

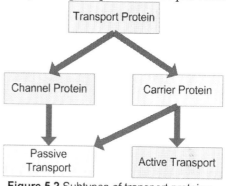

Figure 5.2 Subtypes of transport proteins

Carrier protein. This is any transport protein that alternates between two shapes to facilitate the transportation of an entity across a membrane. Carrier proteins may facilitate passive transport or active transport. In addition, they play a role in cotransport discussed further on in this chapter.

Fascinating Fact! About three billion water molecules can pass through a single aquaporin in a cell membrane. This means, for example, that 3 billion water molecules can enter or leave the aquaporin per second.

A special kind of diffusion across a membrane is called osmosis, which we explore next.

5.5 Osmosis: Water Balance of Cells without Walls

Osmosis is the process in which water diffuses across a selectively permeable membrane; the water moves from the region where its concentration is higher to the region where its concentration is lower.

Some concepts related to osmosis are discussed in the following.

Tonicity. Tonicity is the ability of a solution to gain or lose water. It is influenced only by solutes that cannot pass through a membrane. Tonicity determines the behavior of a cell in a specific solution. Both membrane permeability and solute concentration determine tonicity.

Isotonic. An isotonic solution is a solution that causes no net flow of water through a cell membrane when the cell is immersed in it. This is because the solute concentration is the same inside and outside the cell. Water may still move across the membrane, but it moves at the same rate in both directions.

Hypertonic. A hypertonic solution is a solution that has a higher solute concentration than that of the cell so that the cell loses water when the cell is immersed in the solution. A cell in a hypertonic environment shrivels or shrinks due to loss of water.

Hypotonic. A hypotonic solution is a solution that has a lower solute concentration than that of the cell so that the cell gains water when the cell is immersed in the solution. A cell in a hypotonic environment swells due to gaining water and may rupture.

Osmoregulation. Osmoregulation is the control of solute concentration and therefore water balance in a cell. Species that have cells without walls and live in hypertonic or hypotonic environments have evolved this trait through adaptation.

Plasmolysis. This is a phenomenon in which the plasma membrane pulls away from the cell wall as the cytoplasm of a walled cell shrinks due to the loss of water to the hypertonic environment.

During osmosis, it is water that moves across the membrane from a lower concentration of the solutes to a higher concentration. Ions also move across the membranes and move down the electrochemical gradient. This motion of ions is critical to the functioning of cells. So, let us explore the electrochemical gradient further.

5.6 Membrane Potential and Electrochemical Gradient

Some entities in cells such as ions have electric charges. It turns out that the inside of a cell is electrically negative overall and the outside is electrically positive overall, which gives rise to electric potential or voltage across the plasma membrane This electric potential or voltage is referred to as membrane potential. The magnitude of this membrane potential or voltage ranges from 50 to 200 millivolts (mV). In the following, we describe some concepts that constitute the electrochemical gradient across the membrane.

Electric force. Membrane potential acts like a battery. It is a source of energy which influences the traffic of charged entities across the membrane. Where there is energy, there is force. This potential or force favors the passive transport of cations (positive ions) into the cell and anions (negative ions) out of the cell because the inside of the cell is negative and the outside is positive, and there is an attractive force between opposite charges and a repulsive force between like

Molecular and Cellular Biology

charges. This electric force and the resulting motion only apply to charged entities such as ions.

Chemical force. This is the concentration gradient which applies to both charged and uncharged entities. This force facilitates movement of entities from where they are more concentrated to where they are less concentrated.

Electrochemical gradient or force. The combined effect of electric and chemical forces creates a net force or gradient called the electrochemical gradient which drives the diffusion of charged entities across the membrane.

Electrogenic pump. The transport protein that generates electric potential (voltage) across a membrane is called an electrogenic pump. This protein generates the potential across the membrane as it transports ions actively across the membrane against the concentration potential.

Electrogenic proteins constitute biochemical machines such as the proton pump and the sodium-potassium pump, which are discussed next.

5.7 Sodium-Potassium Pump

The sodium-potassium pump is an example of active transport. It consumes energy in the form of ATP molecules to transport ions against the concentration gradient. In an animal cell where this pump operates, the sodium ion concentration is higher outside the cell, and the potassium ion concentration is higher inside the cell. Yet, the pump transports sodium ions out of the cell and potassium ions into the cell. Here is the process as illustrated in Figure 5.3:

1. **Sodium binding.** The shape of the protein is such that a sodium ion (Na^+) in the cytoplasm of the cell binds to the protein.
2. **Phosphorylation.** The binding of the sodium ion stimulates phosphorylation, which is the addition of a

phosphate group to the protein. Phosphorylation is facilitated by consuming energy in the form of an ATP molecule.

3. **Release of Na^+.** Phosphorylation causes a change in the shape of the protein so that it is no longer able to hold the sodium ion, which as a result is released to the extracellular fluid.

4. **Potassium binding.** The new shape of the protein is such that a potassium ion (K^+) in the extracellular fluid binds to the protein.

5. **Dephosphorylation.** The binding of the potassium ion causes the loss of a phosphate group, a process called dephosphorylation.

6. **Release of K^+.** Dephosphorylation restores the original shape of the protein so that it is no longer able to hold the potassium ion, which as a result is released to the cytoplasm. The shape of the protein is again suitable to capture another sodium ion from the cytoplasm, and the cycle is repeated.

Note that the protein oscillates between two shapes in one cycle. The sodium-potassium pump is critical in the proper functioning of cells such as the communication task in nerve cells and the proper flow of many molecules such as glucose in the body.

So far in this chapter, we have explored active transport and passive transport. However, there are biological machines in some cells that perform a combination of active and passive transports called cotransport.

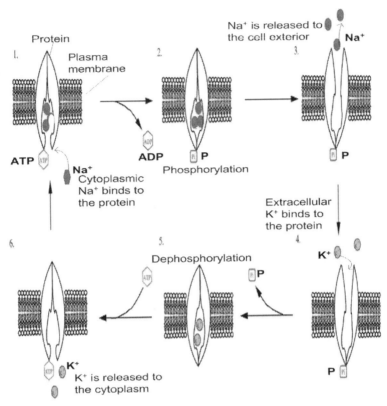

Figure 5.3 Illustration of the sodium-potassium pump.

5.8 Cotransport

Cotransport is the phenomenon in which diffusion, that is the spontaneous or passive transport of one type of entity along or down its gradient, is facilitated by the active transport of another kind of entity against its own gradient. In other words, it is a coupling of passive transport with active transport. An example of cotransport is the proton pump protein coupled with a carrier protein. The process, which is illustrated in Figure 5.4, is described in the following:

1. An active transport protein called a proton pump, which is driven by the energy of the ATP molecules, transports hydrogen ions (H^+) through the membrane from the cytoplasm to the extracellular fluid. This is an electrogenic pump that works against the electrochemical gradient.

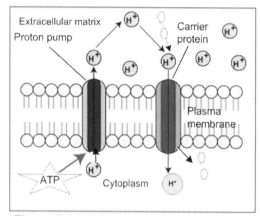

Figure 5.4 An illustration of cotransport.

2. This transport creates (or increases) the electric potential in the membrane: net negative charge in the cytoplasm and net positive charge in the extracellular fluid.

3. Due to this potential, some protons (hydrogen ions) spontaneously (passively) tend to come back into the cytoplasm.

4. Other molecules such as amino acids and sugars piggyback with hydrogen ions from the extracellular fluid into the cytoplasm. A carrier protein that facilitates this one-way trip is called a cotransporter. The cotransporter is separate from the proton pump, and its transport direction is opposite to that of the proton pump.

> **Proton pump versus sodium-potassium pump.** Whereas the sodium-potassium pump is the main electrogenic pump in animals, the proton pump is the main electrogenic pump in bacteria, fungi, and plants, which actively transport protons (hydrogen ions, H^+) across the plasma membrane from the cytosol to the extracellular environment.

Water, other small molecules, and ions enter and leave the cells by diffusion or pumping as discussed so far. However, cells also deal with bulk material such as food that we eat. How does this type of transport happen? Let us explore.

5.9 Bulk Transport Through Cellular Membranes

Cells extract energy from the bulk food that we eat. These large particles or molecules such as polysaccharides and proteins pass through the membrane into the cell generally by using mechanisms that involve packaging them in vesicles. These mechanisms are also used to export bulk material out of the cell. These mechanisms require energy just like active transport and fall into two broad categories: exocytosis and endocytosis. They are described in the following.

Exocytosis. This is the process through which biological molecules are secreted out of the cell by packaging them in vesicles and then fusing the vesicles that carry them with the plasma membrane. For example, consider a transport vesicle that is carrying a protein molecule, which has budded from the Golgi apparatus and has arrived at the plasma membrane after moving along the microtubules of the cytoskeleton. The vesicle membrane fuses into the plasma membrane because the lipid molecules of the two membranes rearrange themselves when the two membranes come in contact with each other. Due to this fusion, the outer face of the vesicle becomes part of the inner face of the plasma membrane, and therefore the molecules in the vesicle are spilled (or secreted) into the extracellular fluid. Consider another example. Some cells in the pancreas make a protein called insulin, which is secreted out of the cell into the blood by exocytosis.

Endocytosis. This is the process by which a cell takes in large molecules and particulate matter by packaging it in vesicles made out of the plasma membrane of the cell. Here is the process. A small portion of the plasma membrane sinks in to make a pocket for the incoming material. Subsequently, the cell engulfs the material into the pocket, and the pocket deepens and pinches in to form a separate sac called a vesicle that now contains the incoming material. There are three subtypes of endocytosis described in the following.

Phagocytosis. In this process, which is also called *cell eater*, the cell engulfs food or other solid particles. The vesicle used in this method is large enough to be considered a vacuole. The material in the vacuole is then digested when the vacuole fuses with a lysosome. Recall that a lysosome contains hydrolytic enzymes that facilitate digestion. Also by using this method, a bacterium can engulf another bacterium. Some cells use it to engulf nutrients, but it is mostly a cell-eating process. Immune systems use it to remove (destroy) pathogens and cell debris.

Pinocytosis. In this process, which is also called *cell drinker*, the cell takes in particles or molecules by gulping droplets of extracellular fluid. The vesicles used in this process are tiny as opposed to the large vesicles used in phagocytosis. Pinocytosis is not selective or specific about taking in the substance because all the solutes come in with the fluid that it takes in.

Receptor mediated endocytosis. In this process, the cell takes in bulk quantities of specific molecules by using the receptor sites of the proteins in the plasma membrane. Only the molecules that can bind to those receptor sites are taken in. After the material in the vesicle is delivered inside the cell, the same vesicle transports the receptor proteins back to the plasma membrane. This method is commonly used by cells to uptake specific substances they need. For example, in human cells, cholesterol molecules are taken in by using receptor mediated endocytosis. These molecules then may be used to synthesize some structures such as membranes and steroids.

> **Note.** A principle of conservation is in action during the processes of endocytosis and exocytosis. The incoming material takes some membrane away from the plasma membrane in the form of a vesicle, and the outgoing material adds membrane to the plasma membrane as the vesicle fuses with it. These processes conserve the membrane material in the plasma membrane.

5.10 In a Nutshell

- Cellular membranes are largely composed of phospholipids and proteins.
- The selective permeability of the plasma membrane, which spontaneously results from its structure, is fundamental to life for two reasons: it enables the cell to exchange chemicals with its environment in a discriminative way, and it allows the cell to maintain an internal environment that is different from its outside environment.
- The internal membranes of a cell facilitate compartmentalized specializations inside the cell.
- In order to work properly, membranes must have the right degree of fluidity.

Cell Membrane: The Supermolecule of Life

- Cholesterol in a membrane works as a buffer that resists a change in the fluidity of the membrane due to a change in temperature.
- Both diffusion and facilitated diffusion are forms of passive transport because no energy is expended, even though proteins accelerate the transport in facilitated diffusion.
- Active transport across cellular membranes is transport in which energy is expended because movement happens against the electrochemical gradient.
- Bulk material transports across the plasma membrane from inside to outside the cell by exocytosis and from outside to inside the cell by endocytosis.

5.11 Review Questions

1. The major structural components of the plasma membrane are ____.

   A. proteins and phospholipids

   B. proteins and carbohydrates

   C. proteins and cholesterol

   D. cellulose and phospholipid

2. What role does cholesterol play in cellular membranes?

   A. It increases their fluidity.

   B. It decreases their fluidity.

   C. It stabilizes them.

   D. It helps one cell communicate with other cells.

3. Which of the following is a true statement about phospholipids in a cellular membrane?

   A. They move faster laterally than protein molecules because they are larger than protein molecules.

   B. They move laterally by switching positions with their neighbors very quickly.

   C. The hydrophilic tails of phospholipids in one layer are attracted to the hydrophilic tails in the other layer, so the hydrophilic tails from both layers are packed in the interior of the membrane.

   D. The phospholipid tails are packed in the interior of the membrane because if they project into the cytosol or into the extracellular environment, they would dissolve in the aqueous environment.

4. Phospholipids with unsaturated hydrophobic tails keep a cellular membrane from freezing because ____.

   A. most of the bonds in the unsaturated hydrocarbons are single bonds

   B. double bonds in the unsaturated hydrocarbons act as kinks that prevent tight packing of hydrocarbons and hence phospholipid tails in the membrane

   C. hydrophobic tails keep the hydrophilic heads of the two lipid layers away from each other

   D. unsaturated tails are not attracted to the aqueous environment of cytosol and the extracellular matrix

5. The rapid flow of water through cellular membranes is facilitated by ____.

A. the small size of the water molecules
B. the polarity of the water molecules
C. proteins called aquaporins
D. proteins called glycoproteins

6. Which of the following must be amphipathic to perform its function in a cellular membrane?

   A. integral protein
   B. phospholipid
   C. peripheral protein
   D. phospholipid tails
   E. A. and B.

7. Which of the following has a helical structure and may span across a cellular membrane multiple times?

   A. nucleic acid
   B. transmembrane protein
   C. peripheral protein
   D. aquaporin

8. The passage of water through aquaporins in a cellular membrane is called ____.

   A. diffusion
   B. osmosis

Molecular and Cellular Biology

C. facilitated diffusion

D. passive transport

E. all of the above

9. What is true about cotransport through a cellular membrane?

    A. It is passive transport.

    B. It is active transport.

    C. It is a coupling of passive and active transport.

    D. It is another name for proton pump transport.

10. Endocytosis is ____.

    A. more difficult in plants than in animals

    B. more difficult in animals than in plants

    C. equally difficult in plants and animals

    D. none of the above

5.12 Answer Key

1. A.	6. E.
2. C.	7. B.
3. B.	8. E.
4. B.	9. C.
5. C.	10. A.

Notes:

Q2. Cholesterol helps to stabilize membranes by reducing their fluidity at higher temperatures (and hence keeps them intact) and by enhancing their fluidity at lower temperatures (and hence helps them resist solidification).

Q10. Plant cells have walls.

Molecular and Cellular Biology

Chapter 6

Cellular Energy

6.1 Cellular Energy: The Big Picture

A cell, the fundamental building block of life, is a microscale chemical factory in which thousands of different chemical reactions take place. Through these reactions, matter and energy are transformed into each other, and one form of energy is transformed into another form. The sum total of these chemical reactions in the cell is called *metabolism*. In Chapter 1, we discussed how new *emergent properties* appear at each organizational level due to the interaction of the components within each organizational level. Metabolism is an emergent property that appears at a cellular level due to the interaction between molecules, the components of the cell.

During the digestion of food, complex molecules are broken down into smaller molecules, and energy is extracted in the process in the form of ATP (adenosine triphosphate) molecules, which are used by the cell for its work. This process of breaking down complex molecules into simpler ones is called *catabolism*. On the other hand, part of cellular work is to use energy in ATP molecules to synthesize complex molecules such as carbohydrates and proteins from simpler molecules. This process is called *anabolism*. Anabolism and catabolism constitute metabolism.

Molecular and Cellular Biology, by Paul Sanghera
Copyright © 2015 Infonential.

Molecular and Cellular Biology

> **Energy Matter Equivalence.** Mass, m, and energy, E, are equivalent and interconvertible according to Einstein's famous equation
>
> $$E = mc^2$$
>
> where c is the speed of light through a vacuum, a constant; m is the relativistic mass; and E is the energy.

So, in the cell, catabolic reactions make ATP molecules, and anabolic reactions spend them. In other words, the ATP molecule is the currency for cellular energy. It is composed of an organic phosphate molecule called *adenosine* covalently attached to a string of three phosphate groups, as shown in Figure 6.0:

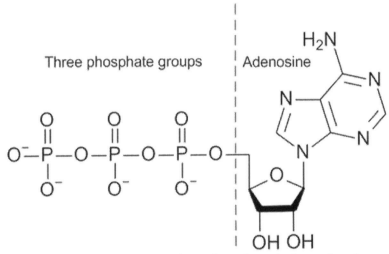

Figure 6.0 ATP: The adenosine phosphate molecule

In anabolic reactions, the ATP molecule releases energy by reacting with a water molecule and transforming into an ADP (adenosine phosphate) molecule:

$$ATP + H_2O \rightarrow ADP + P_i + Energy \quad (6.1)$$

In Equation 6.1, P_i stands for an *inorganic phosphate*, a phosphate group that ATP has lost along with energy in order to turn into ADP. This reaction is an example of hydrolysis.

The reverse is also true: In catabolic reactions, ADP combines with P_i to turn into ATP by consuming some energy:

$$ADP + P_i + Energy \rightarrow ATP + H_2O \quad (6.2)$$

This reaction is an example of dehydration. Through these two chemical reactions, ATP links the two great processes of metabolism: anabolism and catabolism.

In this chapter, we will explore processes that run metabolism, the transformation of matter and energy. When we talk about energy, matter, and their transformation, we cannot escape physics, but we will keep it to a minimum.

6.2 Physical Laws at Work in Metabolism

As mentioned in Section 6.1, metabolism is the sum total of all the chemical reactions inside an organism including reactions that make the catabolic and anabolic pathways. These pathways manage the energy and material resources of the cell. Recall that cells are the fundamental structural and functional units of life. This means metabolism occurs at a cellular level. Energy production and energy transfer are central to metabolism, and this production and transfer occurs according to physical quantities and laws, some of which are depicted in Figure 6.1 and discussed in the following:

Energy. Energy is the capacity to perform work or to cause change. All forms of energy can be broadly categorized into two forms, potential energy and kinetic energy, which are discussed in the following:

- **Potential energy.** This is the energy that matter possesses because of its position or structure. Some examples of potential energy include the energy that a ball stuck at the top of a tree possesses due to its position and energy in the form of calories in a glucose molecule.

- **Kinetic energy.** This is the energy due to the motion of an object. For example, when the ball stuck at the top of the tree is set free, it begins moving toward the ground due to gravity, and its potential energy starts converting into kinetic energy.

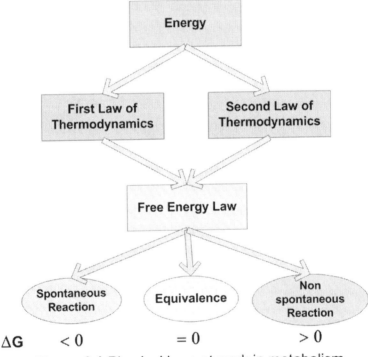

Figure 6.1 Physical laws at work in metabolism

Conservation of Matter and Energy: The First Law of Thermodynamics. Energy of the universe is conserved: it can neither be created nor destroyed; it can only be transformed from one form into another. Because energy and matter are interconvertible, it is actually the sum of energy and matter that is conserved. Also because energy and matter are interconvertible, when we refer to energy, we also mean matter and vice versa. Scientists often use the terms *matter* and *energy* interchangeably. An equivalent statement of the energy conservation law is that the energy in a process is conserved; that is, energy gained by a system under study is equal to the energy lost by its environment. In other words, the energy of the universe in any process remains conserved (unchanged in amount).

Second Law of Thermodynamics. Entropy, a degree of randomness and disorder, of the universe stays the same or increases in each transfer or transformation of energy. Even though energy cannot be created, it can be transferred and transformed. This is how living organisms get their energy. For example, our electric energy is obtained by transforming the energy of fossil fuel such as coal, crude oil, or natural gas; the mechanical energy of falling water; or the energy of the atomic nuclei. As another example, plants transform energy from sunlight into carbohydrates that we eat, and the energy from these carbohydrates is converted by our cells into energy molecules called adenosine triphosphate (ATP), which cells can then use to perform bodily functions. However, it turns out that not all energy can be transferred and cycled in a usable form. During these energy transfers and transformations, some energy is lost as unusable energy, which adds to the disorder or entropy of the universe. The second law of thermodynamics states that in each energy transfer, the entropy of the universe either remains the same or increases, but it never decreases.

Gibbs Free Energy Law. The two laws of thermodynamics discussed in this section refer to the energy of the universe, a given system plus its environment. However, Gibbs Free Energy Law can be applied to a given system without considering its environment by using the following equation:

$$\Delta G = \Delta H - T\Delta S \quad (6.3)$$

In Equation 6.3, ΔH is the change in total energy of the system (negative if the system has lost energy), ΔS is the change in the system's entropy, and T is the temperature of the system in Kelvin (K) units. The quantity ΔG is referred to as the free energy of the system.

Any chemical reaction can be considered as a system. In a chemical reaction, there are three possibilities for change in free energy, ΔG, which are discussed in the following:

$\Delta G < 0$. In this case, the reaction happens spontaneously because the system loses enthalpy (free energy). A reaction which loses (or releases) free energy is called an *exergonic reaction*.

$\Delta G = 0$. In this case, the reaction is in equilibrium. In a chemical reaction it means the reaction in forward and backward directions happens at equal rates. This is the state in which a system has maximum stability.

$\Delta G > 0$. In this case, the reaction happens nonspontaneously because it gains energy or requires energy before it could occur. A chemical reaction that consumes energy from its surroundings is called an *endergonic reaction*.

The theme behind these rules is that a system tends to become stable by acquiring a state of minimal free energy, G:

$$G = H - TS \quad (6.4)$$

We define energy laws in terms of a system. A system can be a part of an organism, a community, or the whole ecosystem.

Not only must chemical reactions occur in order to maintain life, but they must also occur at certain rates. A special type of protein in the cells called enzymes makes these reactions possible.

6.3 Energy, Enzymes, and Chemical Reactions

In a chemical reaction, some existing bonds between atoms break, and some new atomic bonds are formed. This results in making new molecules from the reacting molecules. Regardless of whether a reaction is exergonic or endergonic, it needs a certain threshold amount of energy to occur completely. This is because the molecules need to go into an unstable state before the atomic bonds keeping them together can be broken. In order to go into that unstable state, the molecules must absorb a certain amount of energy from their surroundings. This minimum amount of energy that the molecules must absorb before the reaction can be completed is called activation energy. As illustrated in Figure 6.2, biological catalysts called enzymes lower the activation energy, and this way accelerate the reaction at a given temperature.

Enzymes are a type of protein. There are thousands of enzymes in each cell because a given enzyme supports only a specific reaction. The reactants on which an enzyme operates are called *substrates*. The part of the enzyme interacting with the substrate is called the *active site* of the enzyme.

Let us take a closer look at Figure 6.2. The initial energy of the reactants is E_i. Before the reaction can occur, they must reach the energy level of E_h in the absence of an en-

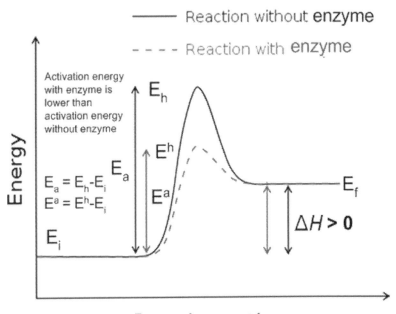

Reaction path

Figure 6.2 Effect of an enzyme on a chemical reaction

zyme or an energy level of E^h in the presence of an enzyme. As Figure 6.2 clearly shows, the activation energy $E^a = E^h - E_i$ in the presence of an enzyme is less than the activation energy $E_a = E_h - E_i$ in the absence of an enzyme. After reaching the energy peak (E^h), the molecules lose some energy in order to get stabilized as products at energy E_f in this example.

Note from Figure 6.2 that an enzyme does not change the ΔH or ΔG of the reaction.

> **Think About It!**
> Is the reaction represented in Figure 6.2 exergonic or endergonic?
>
> **Answer:**
> Because the energy of the product, E_f, is higher than the energy of the reactants, E_i, this is an example of an endergonic reaction.

Matter-energy transformation happens at all complex levels of life from cells to ecosystems.

6.4 Energy Cycle in an Ecosystem

Figure 6.3 illustrates the big picture of energy (or energy/matter) cycle in an ecosystem. Energy flows into an ecosystem in the form of sunlight. Cells in some organisms

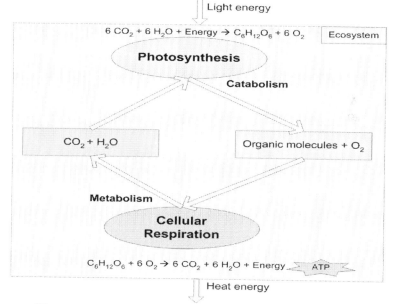

Figure 6.3 Energy-matter flow cycle in an ecosystem

Molecular and Cellular Biology

called photoautotrophs use this light to produce organic molecules and oxygen. Life depends on these molecules and oxygen. Some energy leaves the ecosystem in the form of heat. Here are some energy proceeses that happen in an organism at a cellular level:

Photosynthesis. This is a metabolic pathway used by photoautotrophs to convert light energy, carbon dioxide, and water into organic molecules such as carbohydrates and oxygen. Photosynthesis occurs in plants, algae, and certain prokaryotes. A metabolic pathway is a series of biochemical reactions that either build a complex molecule from simpler ones or breaks down a complex molecule into smaller (simpler) ones.

Cellular respiration. This is the process of breaking down organic molecules in order to produce cellular energy in the form of ATP molecules.

Catabolism. This is a metabolic pathway (set of chemical reactions) that releases energy by converting complex molecules to simpler ones. Cellular respiration is an example of catabolism.

Anabolism. This is a metabolic pathway that uses energy to synthesize a complex molecule from simpler ones. Photosynthesis is an example of anabolism.

Catabolism and anabolism are two pathways that constitute metabolism, which we discuss next.

6.5 Metabolism: Making and Breaking of Molecules

Metabolism is the process of transformation of matter and energy by making and breaking complex molecules. Figure 6.4 presents a bird's eye view of metabolism. A living cell is never in equilibrium. It takes in material, breaks it down into simpler molecules, release energy, and then uses that energy to perform work, which includes making more complex molecules from simpler molecules. Metabolism and its major processes are depicted in Figure 6.4 and described in the following:

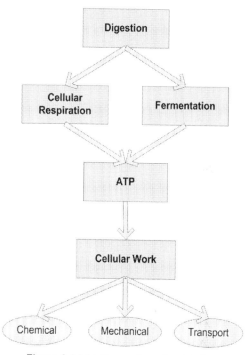

Figure 6.4 A bird's eye view of metabolism

1. **Digestion.** This is the process of breaking down food into molecules that are small enough for the cells to take in and process. Chewing is the first step in digestion and is called mechanical digestion, and mechanical digestion is followed by chemical digestion.

2. **Cellular respiration.** This is the process of breaking down organic molecules in order to produce cellular energy in the form of ATP molecules. The process is an

example of *catabolism*. Respiration is of two types: *aerobic respiration*, which uses oxygen to produce ATP, and anaerobic respiration, which does not use oxygen. Humans use aerobic respiration.

3. **Fermentation.** This is a catabolic process of breaking down organic molecules to produce ATP. This process is less involved than cellular respiration and produces a smaller number of ATP molecules. It also produces other products such as ethyl alcohol or lactic acid.

4. **ATP (adenosine triphosphate).** The product of cellular respiration and fermentation, this is a nucleoside molecule that releases energy when its phosphate bonds are hydrolyzed. The energy of this molecule is used to perform cellular work.

5. **Cellular work.** A cell performs three kinds of work: chemical, mechanical, and transport.

    - **Chemical work.** These are running nonspontaneous reactions called endergonic reactions because energy is spent to run them. An example is the set of reactions that synthesize complex molecules such as proteins and lipids from simpler molecules through anabolism.

    - **Mechanical work.** Examples of mechanical work are the movement of molecules such as chromosomes and proteins and the contraction of muscle cells.

    - **Transport work.** Examples of transport work include the pumping of molecules across membranes that is also known as active transport.

    Metabolism includes catabolism and anabolism discussed next.

> Caution! Chemical digestion continues in the stomach, but it actually begins in the mouth. Your mouth creates saliva, which contains amylase, an enzyme that breaks down complex carbohydrates into smaller ones.

6.6 Catabolism: Breaking of Complex Molecules

Carbohydrates, fats, and proteins can all be used to obtain energy. Complex carbohydrates are broken down (hydrolyzed) in the digestive tract into glucose, which then goes through hydrolysis inside the cell. Proteins are broken down into amino acids. Many amino acids from proteins are then used in anabolism to synthesize other proteins needed by the organism. However, as shown in Figure 6.5, amino acids present in excess can be used to generate energy after removing the amino group, a process called *deamination*. There are three major stages of catabolism:

1. Glycolysis. This is a process that splits one glucose molecule into two pyruvate molecules. Glycolysis serves as the starting point for cellular respiration and fermentation, and it occurs in almost all living cells. It occurs in the cytosol of a cell, and its net energy yield is 2 ATP molecules for one glucose molecule.

2. Citric acid cycle. Also called the Krebs cycle, it is the second stage of catabolism which completes the breakdown of glucose by oxidizing pyruvate molecules to CO_2. This occurs in the cytosol of a prokaryotic cell and within a mitochondrion of a eukaryotic cell. The net energy yield of the citric acid cycle is 2 ATP per glucose molecule.

Molecular and Cellular Biology

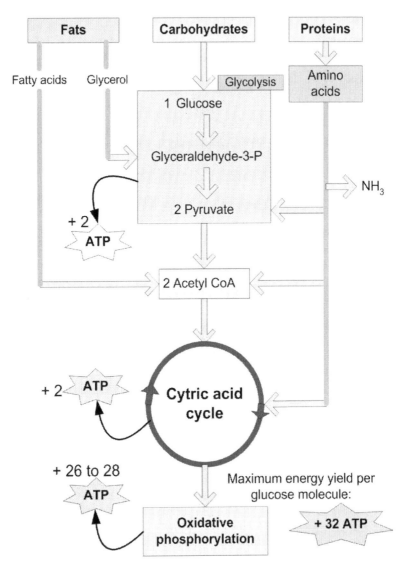

Figure 6.5 Catabolism of molecules from food

3. **Oxidative Phosphorylation.** This is the third major stage of catabolism during which a large amount of cellular energy is generated in the form of ATP by using the energy derived from the redox reactions of an electron transport chain. Energy lost by the electrons falling down the electron transport chain pushes protons (H^+) out of the membrane and thereby creates a proton gradient across the membrane. The energy stored in the proton gradient is used to synthesize 26 to 28 ATP molecules per glucose molecule. In this step, the energy yield per glucose molecule is 26 to 28 ATP molecules. This process of using the energy from the proton gradient to perform cellular work, for example to synthesize ATP in this case, is called *chemiosmosis*.

The maximum net energy yield per glucose molecule during all stages of a respiration cycle is 32 ATP molecules.

There are some variations in catabolism as described in the following:

1. **Cellular aerobic respiration** goes through all the three stages described above.

2. **Fermentation** does not use oxidative phosphorylation, and hence it has no electron transport chain. It uses only glycolysis plus some chemical reactions to regenerate NAD^+. To produce ATP, it uses substrate-level phosphorylation just like glycolysis.

3. **Cellular anerobic respiration**, which occurs in certain prokaryotic organisms, uses all three stages of catabolism including the electron transport chain, but instead of using oxygen at the end of the electron transport chain, it uses some other material such as sulfate ions to accept the electrons.

So, fermentation and anaerobic respiration use catabolism to produce ATP without the use of oxygen. The whole process of converting the biochemical energy of nutrients (carbohydrates, lipids, and proteins) into ATP and releasing waste products is called cellular respiration. It is a reduction-oxidation (redox) process, which means it involves a chemical reaction that is a reduction-oxidation reaction.

6.7 Cellular Respiration as a Reduction-Oxidation Process.

During the process of cellular respiration, energy is released when electrons associated with hydrogen atoms in a sugar molecule travel with the help of carriers to oxygen. The potential energy of the electrons is released in the form of ATP as they fall down the electron transport chain toward more electronegative oxygen. Think of transforming the potential energy of water in a dam into electricity (electric energy) when it falls down. The general downhill route for the electrons is as follows:

Glucose → NADH → Electron transport chain → Oxygen

The following is the detail of this redox process as illustrated in Figure 6.6:

1. **Electrons embark on a journey.** Hydrogen atoms are stripped from glucose as the glucose is oxidized. Look at each of these hydrogen atoms as an electron with a proton. These electrons travel with their protons, that is, as hydrogen atoms.

2. **Capturing of electrons by NAD^+.** A pair of electrons in the form of two hydrogen atoms meets with a coenzyme NAD^+ and reduces it to NADH:

    $2H + NAD^+ \rightarrow NADH + H^+$ (6.5)

Cellular Energy

3. **Shuttling of electrons by NADH.** NADH shuttles the electrons associated to it to the top (higher energy end) of the electron transport chain, which consists of proteins and other molecules built into the inner membrane of a mitochondrion in a eukaryotic cell and the plasma membrane of prokaryotes that perform anaerobic respiration.

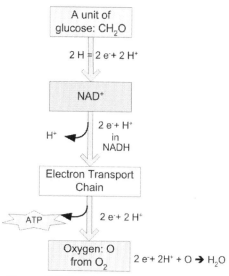

4. **The fall of the electrons.** In a series of steps, electrons fall from a

Figure 6.6 Cellular respiration as a redox process

state of high potential energy to meet highly electronegative oxygen, which has a great affinity for electrons. In this fall, the lost potential energy of the electrons is released in the form of ATP molecules.

Although it happens in several steps, the end result is that a glucose molecule reacts with six oxygen molecules to produce six molecules of carbon dioxide, six molecules of water, and more than 30 molecules of ATP. Here is the overall redox reaction:

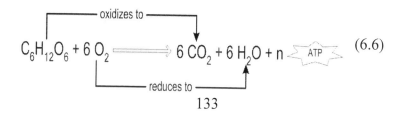

(6.6)

A redox (oxidation-reduction) reaction is a chemical reaction in which an atom completely or partially loses one or more electrons (oxidation), and another atom gains those electrons (reduction). Equation 6.3 illustrates a redox reaction because glucose (or the H atoms in it) loses an electron and as a result oxidizes to carbon dioxide, and the oxygen atoms in the oxygen molecules gain those electrons and as a result reduce the oxygen molecule to water.

So, cellular respiration can be looked upon as a redox process with glycolysis as its first step.

6.8 Glycolysis: Energy Production by Splitting Glucose Molecules

Glycolysis means splitting glucose. During the first stage of glycolysis called the energy investment phase, which is depicted in Figure 6.7, two molecules of ATP are consumed

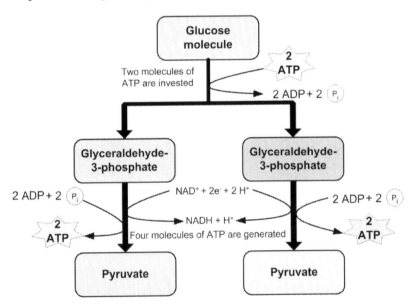

Figure 6.7 Glycolysis producing two molecules of pyruvate (pyruvic acid) from one molecule of glucose

to split a 6-C sugar called glucose into two 3-C sugars, glyceraldehyde-3-phosphate. These two molecules of 3-C sugar are then oxidized, and their bonds are rearranged to turn them into two molecules of pyruvate during the second stage called the energy payoff phase, which yields 4 ATP molecules. So, the net gain of cellular energy is 2 ATP. Here are the chemical reactions involved in glycolysis:

Energy Investment Phase Reaction. Adenosine triphosphate (ATP) produces adenosine diphosphate and inorganic phosphate (P_i):

$$2 \text{ ATP} \rightarrow 2 \text{ ADP} + 2 \text{ P}_i \quad (6.7)$$

where P_i symbolizes an inorganic phosphate group.

Energy Payoff Phase Reaction. ADP combines with P_i to produce ATP:

$$4 \text{ ADP} + 4 \text{ P}_i \rightarrow 4 \text{ ATP} \quad (6.8)$$

$$2 \text{ NAD+} + 4 \text{ e}^- + 4 \text{ H}^+ \rightarrow 2 \text{ NADH} + 2 \text{ H}^+ \quad (6.9)$$

What happened to glucose? This is what happened overall:

$$\text{Glucose} \rightarrow 2 \text{ Pyruvate} + 2 \text{ H}_2\text{O} + 2 \text{ ATP} \quad (6.10)$$

Note the following:

- Six atoms of carbon from the glucose molecule are now in the two molecules of pyruvate; no CO_2 is produced.

- Glycolysis occurs regardless of whether or not oxygen is present.

- Two molecules of pyruvate transport energy go toward the citric acid cycle, and two molecules of NADH transport energy in the form of high energy electrons go to the electron transport chain for the process of oxidative phosphorylation if oxygen is present.

Glycolysis is followed by the citric acid cycle.

Molecular and Cellular Biology

6.9 Citric Acid Cycle: The Big Energy Production Machine

In eukaryotic cells, if oxygen is present, a pyruvate molecule from glycolysis enters a mitochondrion where some enzymes complete the oxidation of glucose, which started during glycolysis as described in the previous section. Figure 6.8 illustrates what happens in the citric acid cycle.

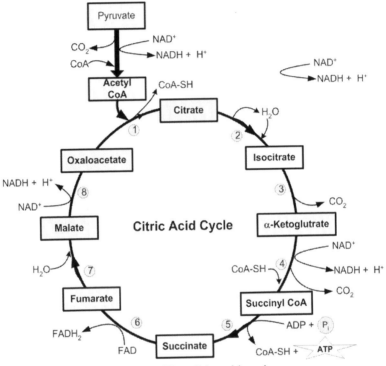

Figure 6.8 The citric acid cycle.

1. Acetyl CoA, which is produced from the oxidation of pyruvate, contributes its 2-C acetyl group to oxaloacetate, resulting in the production of citrate.

2. A molecule of water is added to and removed from the citrate to convert it to its isomer called isocitrate.

3. NAD$^+$ oxidizes isocitrate and reduces itself to NADH in the process. After losing a CO$_2$ molecule, the oxidized isocitrate converts into α-ketoglutarate.

4. Another round of oxidization by NAD$^+$ and the release of a CO$_2$ molecule turns α-ketoglutarate into succinyl CoA with the help of CoA-SH.

5. Phosphorylation produces an ATP molecule, converting succinyl CoA to succinate after releasing CoA-SH.

6. FAD oxidizes succinate (with two negative charges) to fumarate by using two protons (H$^+$) released in the previous steps. In this process, FAD itself is reduced to FADH$_2$.

$$\text{Succinate}^{2-} + FAD + 2 H^+ \rightarrow \text{Fumarate} + FADH_2 \quad (6.11)$$

7. The addition of a water molecule to fumarate rearranges its bonds to give rise to malate.

8. NAD$^+$ oxidizes malate to oxaloacetate, reducing itself to NADH in the process. With the reappearance of oxaloacetate, we are ready for another round with the arrival of an acetyl CoA.

Note that the citric acid cycle in one turn produces one molecule of ATP through substrate-level phosphorylation, 3 NADH, and a FADH$_2$. The NADH and FADH carry electrons to the electron transport chain, the next step of respiration.

$$\text{Acetyl CoA} + 3\ NAD^+ + FAD + ADP + P_i \rightarrow 3\ NADH + FADH_2 + 2\ CO_2 + ATP \quad (6.12)$$

The equation does not include everything; it's here to give you some feel of what happens in the cycle. Carbon skeletons for amino acid biosynthesis are supplied by intermediates of the citric acid cycle such as α-ketoglutrate.

Our cells use respiration to break down food molecules that we eat to produce cellular energy in the form of ATP molecules. But where does food ultimately come from? The answer is photosynthesis in plants.

6.10 Photosynthesis: The Ultimate Source of All Food on Earth

A group of organisms called photoautotrophs make their own food (sugar and other molecules) from light such as sunlight, water, and carbon dioxide by using a process called photosynthesis. Photoautotrophs include algae, plants, cyanobacteria, and some unicellular protists such as euglena.

During the process of photosynthesis, light, water, and carbon dioxide are converted into oxygen and sugar. The overall reaction is as follows:

$6 CO_2 + 6 H_2O + \text{Light energy} \rightarrow C_6H_{12}O_6 + 6 O_2$ (6.13)

As illustrated in Figure 6.9, photosynthesis takes place in two major stages: light reactions and the Calvin cycle. These stages are described in the following:

Light reactions. These reactions

- Convert light energy into the chemical energy of ATP and NADPH
- Obtain electrons for the electron transport chain by splitting water and releasing oxygen
- Are carried out by molecules in the *thylakoid* membranes.

Cellular Energy

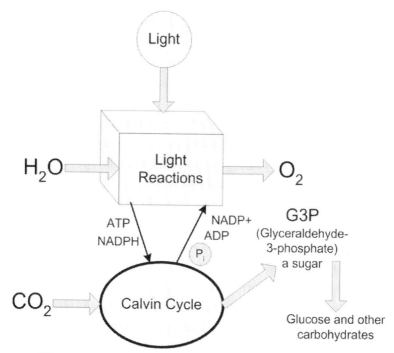

Figure 6.9 Overview of photosynthesis

Calvin cycle reactions. These reactions

- Convert CO_2 into G3P sugar by using ATP and NADPH supplied by light reactions.
- Input ADP, inorganic phosphate P_i, and $NADP^+$, back into light reactions
- Are carried out in the *stroma*, the aqueous fluid of the chloroplast.

Molecular and Cellular Biology

Chloroplast contains thylakoids and stroma. The site of photosynthesis in plants is a double membrane-bound organelle called chloroplast, which is illustrated in Figure 6.10. The dense fluid inside the double membrane of a chloroplast is called stroma. Thylakoids are the membranous sacs in the stroma. It is

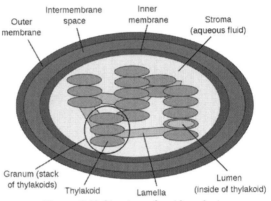

Figure 6.10 Structure of a chloroplast

the membranes of thylakoids that contain the molecular machinery to perform photosynthesis, which converts light energy to chemical energy. Here are some other related terms:

Mesophyll. Chloroplasts are mainly found in the cells of plant tissue called *mesophyll*, the tissue in the interior of a leaf.

Stomata. These are microscopic pores that function as doorways for CO_2 to enter the leaf and for O_2 to exit.

Chlorophyll. This is the name of the molecule that absorbs light and is responsible for the green color of a leaf. Chlorophylls are located within chloroplasts in the thylakoid membranes. The molecules that absorb light are called pigments. Thus, chlorophylls are also called green pigments.

Now that we have covered the main parts of its machinery, let us explore more details of photosynthesis. Here are the main steps of photosynthesis as illustrated in Figure 6.11:

Cellular Energy

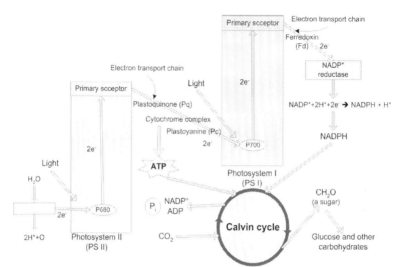

Figure 6.11 Sugar production from water and carbon dioxide in photosynthesis

1. A photon of light hits an electron in a chlorophyll molecule in a light harvesting complex of the chloroplast, boosting the electron to a higher energy state. This electron falls back to its ground state by transferring its energy to an electron in a neighboring chlorophyll molecule, which in turn raises to an excited state. This process continues until the electron in a P680 chlorophyll molecule raises to an excited state.

2. The excited electron in a P680 molecule is captured by the primary electron acceptor resulting in an electron hole (a + electric charge) in the P680 molecule, giving rise to $P680^+$.

3. With the help of an enzyme, a water molecule is split into two electrons, two protons (H^+), and one oxygen (O) atom. Two electrons fill the holes of two $P680^+$ molecules created by two electrons excited by two photons and captured by the primary acceptors shown in

141

Molecular and Cellular Biology

Figure 6.11. Two oxygen atoms from two water molecules combine to form an oxygen molecule.

4. Electrons captured by the primary acceptor go downhill on the electron transport chain from PS II to PS I, generating ATP. This ATP is created from an ADP by adding to it a phosphate group powered by the energy that originally came from the light capture. For this reason, the process that generates this ATP is called *photophosphorylation*.

5. The electrons coming down the electron transport chain fill the holes in the P700 chlorophyll molecules. These holes are created when photons hit the chlorophyll molecules and the excited electrons of P700 are captured by the primary acceptor of PS I as shown in Figure 6.11.

6. Electrons captured by the primary acceptor of PS I go down an electron transport chain. The two electrons coming down the chain and a proton (H^+) reduce an $NADP^+$ to NADPH.

7. This NADPH and ATP (from the first electron transport chain) become an input to the Calvin cycle.

Chemiosmosis. The flow of electrons through an electron transport chain generates a proton gradient, a process called chemiosmosis, that leads to the synthesis of ATP by using ADP and a phosphate group.

In a nutshell, light reactions transform light energy into the chemical energy of ATP and NADPH.

The Calvin cycle takes NADPH, ATP, and CO_2 as input, produces sugar as output, and returns $NADP^+$, P_i, and ADP to the light reactions area.

Let us explore the Calvin cycle in more detail.

6.11 Calvin Cycle: The Other Half of the Citric Acid Cycle

Unlike the citric acid cycle, which is catabolic, the Calvin cycle is anabolic because it builds a carbohydrate, G3P, from smaller molecules by consuming energy in the form of ATP. It takes three molecules of CO_2, one at a time, to generate one molecule of G3P. As illustrated in Figure 6.12, there are three phases of the Calvin cycle described in the following:

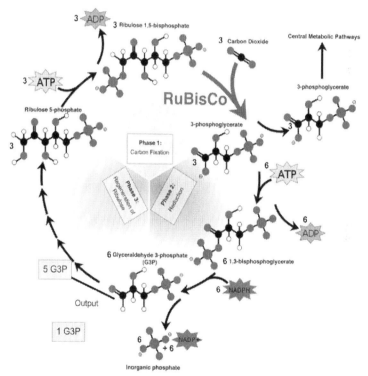

Figure 6.12 The Calvin Cycle.

Phase 1: Carbon Fixation. Carbon fixation, in general, is the process of incorporating carbon from the inorganic compound CO_2 into an organic compound by a photosynthetic organism such as a plant or a chemoautotrophic prokaryote. The Calvin cycle begins with an enzyme *rubisco* attaching a CO_2 molecule to a 5-C sugar called *ribulose biphosphate* (RuBP). The resulting 6-C intermediate, being highly unstable, immediately splits into two molecules of 3-phosphoglycerate. Three molecules of CO_2 generate six molecules of 3-phosphoglycerate by consuming 6 molecules of ATP.

Phase 2: Reduction. Each molecule of 3-phosphoglycerate generated in Phase 1 transforms into 1,3-biphosphoglycerate after accepting a phosphate group from an ATP molecule. Electrons in NADPH reduce a carboxyl group in a 1, 3-biphosphoglycerate to an aldehyde group by turning the molecule into a G3P, which is the same 3-C sugar that forms in glycolysis as well. For three molecules of CO_2, six molecules of G3P are generated. Only one molecule of G3P leaves the cycle; the other five molecules move into Phase 3 to ultimately regenerate three molecules of RuBP to keep the cycle going.

Phase 3: Regeneration of RuBP. The five molecules of G3P from Phase 2 are converted into three molecules of RuBP after the bonds in the carbon skeleton of the G3P molecules are rearranged in a series of reactions. Three molecules of ATP are consumed to make this rearrangement happen. The three molecules of RuBP are now ready to accept three molecules of CO_2 and begin the cycle again.

In a nutshell, the Calvin cycle reduces carbon dioxide to sugar by using the chemical energy of ATP and NADPH, which are produced in light reactions.

The overall process of transforming light energy into ATP is sometimes called *phosphorylation* because it involves transforming ADP to ATP by capturing a phosphate group.

A careful reader must have noticed that the electron transport chain plays a role in both processes of ATP synthesis: respiration and photosynthesis. Let us explore it further.

6.12 Electron Transport Chain: What is It?

The electron transport chain is a sequence of molecules such as proteins which transports electrons from the donors to the acceptors and ultimately releases the energy of the electrons in the form of ATP molecules. The electron transport chain gets its name from the fact that electrons are transported from the donor to meet up with oxygen (for example from respiration) at the end of the chain, where protons (H^+) combine with electrons (e^-) to form hydrogen atoms, which in turn combine with oxygen atoms to from water by releasing energy in the form of ATP molecules. This overall electron chain transport reaction, illustrated in Figure 6.13, to produce energy is shown below:

$$2 H^+ + 2 e^- + 1/2 O_2 \rightarrow H_2O + \text{energy (ATP)} \quad (6.14)$$

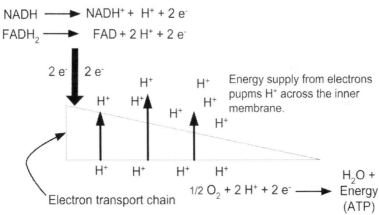

Figure 6.13 Proton gradient created by electron transport chain

Molecular and Cellular Biology

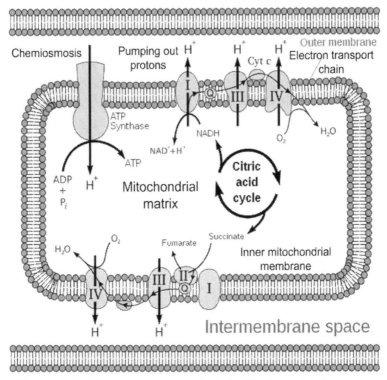

Figure 6.14 Electron transport chain facilitating oxidative phosphorylation in mitochondrion during cellular respiration.

As you have learned in this chapter, the electron transport chain is used in both cellular respiration and photosynthesis to harness energy for transforming ADP into ATP. To accomplish this, both cellular respiration in a mitochondrion and photosynthesis in a chloroplast use the same process called chemiosmosis explained below. As illustrated in Figure 6.14 for mitochondria and Figure 6.15 for chloroplast, the electron transport chain exists in a membrane in which electrons are transported by a series of carriers such as protein molecules, which are progressively more electronegative. This travel of electrons is a series of redox reactions in which electrons lose energy which is used to push protons across the membrane.

Cellular Energy

This way the electron energy is stored as a proton motive force in the proton gradient across the membrane. Subsequently, the protons (hydrogen ions) diffuse down their gradient, and their energy is used to transform ADP into ATP, a process called oxidative phosphorylation in a mitochondrion and photophosphorylation in a chloroplast. The original source of electrons that travel down the electron transport chains is, of

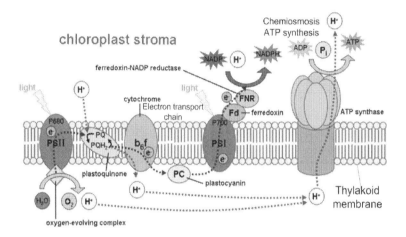

Figure 6.15 Electron transport chain facilitating photophosphorylation in chloroplast during photosynthesis.

course, different in respiration and photosynthesis.

The electron transport chains in both a mitochondrion and a chloroplast use mobile, lipid-soluble carriers such as quinones and mobile, water-soluble carriers such as cytochromes. Furthermore, they contain proton pumps (protein molecules) to push hydrogen ions across the membrane in order to create the proton gradient.

Molecular and Cellular Biology

6.13 Summary: Putting It All Together

Table 6.1 Summary of some important processes carrying out matter energy transformation in the cell.

Process	What It Does	Where It Happens
Glycolysis	First stage of aerobic respiration, anaerobic respiration, and fermentation It produces two molecules of pyruvate and two molecules of ATP from one molecule of glucose.	Cytosol ATP substrate-level phosphorylation
Krebs cycle Also called the citric acid cycle	Second stage of aerobic (and anaerobic) respiration It takes in acetyl-CoA (produced from pyruvate) and generates ATP, NADH, CO_2, and $FADH_2$ and produces 2 ATP molecules per glucose molecule.	Inner membrane of mitochondrion in eukaryotic cells Cytosol of prokaryotic cells ATP produced by substrate-level phosphorylation
Oxidative phosphorylation	Third stage of aerobic respiration The electron transport chain takes in electrons from $FADH_2$ and NADH by oxidizing them to FAD and NAD^+ respectively, which are recycled back to glycolysis and the citric acid cycle. The electron transport chain creates the proton gradient; this gradient energy is transformed into the chem-	Inner membrane of mitochondrion in eukaryotic cells Plasma membrane of prokaryotic cells ATP produced by oxidative phosphorylation and

	ical energy of 26 to 28 ATP molecules.	chemiosmosis
Photosystem II	First stage of photosynthesis A complex of chlorophyll molecules including two molecules of P680 gets energized by light with a wavelength of about 680 nm and as a result sends the electrons from water splitting to the electron transport chain where electrons lose energy to generate ATP.	Thylakoid membrane of chloroplast
Photosystem I	Second stage of photosynthesis It has pretty much the same composition as PSII; it includes a pair of P700 molecules instead of P680. These molecules get energized by absorbing light with a wavelength of about 700 nm and as a result escalate the electrons received from PSII through the electron transport chain to the second electron transport chain. $NADP^+$ is reduced to NADPH when electrons fall through the second electron transport chain.	Thylakoid membrane of chloroplast in eukaryotic cells Or Plasma membrane of some prokaryotic cells
Calvin Cycle	Third stage of photosynthesis By using the energy of ATP from the electron transport chain after PSII and FADH from the electron chain, PSI changes CO_2 to G3P, which is ultimately converted to organic molecules such as carbohydrates. It also produces ADP and $NADP^+$ (by oxidizing ATP and NADPH), which are recycled back to the light reactions to reproduce ATP and NADPH.	The stroma, the dense fluid within the chloroplast Cytosol (cytoplasm) of prokaryotic cells

Molecular and Cellular Biology

6.14 In a Nutshell

Cellular processes maintain life by transforming matter and energy at a molecular level. This transformation is carried out through a set of chemical reactions collectively called metabolism. Here are a few points about this transformation:

- The laws of thermodynamics apply to the universe as a whole, that is, the given system and its surroundings, whereas free energy law applies just to a system such as a chemical reaction.
- Energy stored in organic molecules of food originally comes from the Sun.
- A living cell is not in equilibrium, so it continues to perform work during its lifetime.
- Glycolysis is the first stage of catabolism in both cellular respiration and fermentation.
- Although photosynthesis occurs largely in plants, it also occurs in some protists such as algae and some prokaryotes such as cyanobacteria.
- During photosynthesis, the excitation of the chlorophyll molecule caused by the absorption of light energy triggers a chain of events that leads to ATP production.
- The green color of plants comes from the color pigment molecule called chlorophyll.
- Plants don't just make sugars; they use them too. Therefore, like animals, plants break sugars down to fuel cellular work.
- The oxygen released during photosynthesis comes from water and not from carbon dioxide molecules.
- Within a chloroplast, light reactions occur in the thylakoid membranes, whereas the Calvin cycle occurs in the stroma.

Cellular Energy

- In photosynthesis, electrons for the transport chain originally come from water molecules, whereas in respiration, they come from organic molecules such as glucose.
- Bundle sheath cells in C_4 plants use only PSI and not PSII because they use cyclic electron flow to generate ATP to power the cycle that assimilates CO_2 into organic acids and then releases it to the Calvin cycle.
- The sugar made during photosynthesis in chloroplasts supplies chemical energy for the entire plant and carbon skeletons to synthesize all the major organic molecules in the plant, including carbohydrates, lipids, and proteins.
- Photosynthesis is responsible for the presence of oxygen in the Earth's atmosphere.

6.15 Review Questions

1. Oxygen is used to produce ATP by breaking down complex molecules in the following processes:

    A. fermentation

    B. aerobic respiration

    C. anaerobic respiration

    D. A. and B.

    E. none of the above

2. ATP is produced by using which of the following molecules?

    A. carbohydrates

    B. proteins

    C. lipids

Molecular and Cellular Biology

D. nucleic acids
E. all of the above
F. A., B., and C.

3. What happens to NAD^+ during respiration?

   A. NAD^+ is oxidized to NADH during glycolysis.
   B. NAD^+ is reduced to NADH during the citric acid cycle and glycolysis.
   C. NADH is oxidized to NAD^+ in the citric acid cycle in order to produce ATP.
   D. NAD^+ is broken down to create energy.

4. Which of the following is a true statement about respiration in a eukaryotic cell in the presence of oxygen?

   A. Glycolysis takes place in cytosol, whereas the citric acid cycle takes place in the mitochondrion.
   B. Glycolysis and the citric acid cycle take place in cytosol, whereas oxidative phosphorylation takes place in the mitochondrion.
   C. Glycolysis takes place in the outer membrane of the mitochondrion, whereas the citric acid cycle takes place in the inner membrane of the mitochondrion.
   D. Glycolysis takes place outside of the cell.

5. During glycolysis, the net gain of cellular energy is ____.
   A. 4 ATP
   B. 32 ATP

C. 2 ATP

D. 1 ATP

E. 0 ATP

6. What is the input into the citric acid cycle?

    A. pyruvate

    B. glucose

    C. NADH

    D. acetyl CoA

7. Which of the following is not true?

    A. During glycolysis, ATP is produced by substrate-level phosphorylation.

    B. During the citric acid cycle, ATP is produced by substrate-level phosphorylation.

    C. During fermentation, ATP is produced by oxidative phosphorylation.

    D. Oxidative phosphorylation produces most of the ATP during cellular respiration.

8. Which of the following processes produces ATP from energy that comes from light?

    A. photophosphorylation

    B. substrate-level phosphorylation

    C. oxidative phosphorylation

    D. light reactions

Molecular and Cellular Biology

9. The primary function of the Calvin cycle is to ____.

    A. produce ATP

    B. produce sugar

    C. reduce $NADP^+$ to NADPH

    D. split water to create oxygen

10. In plants, photosynthesis occurs in which organelle?

    A. stroma

    B. thylakoid

    C. chlorophyll

    D. chloroplast

    E. all of the above

6.16 Answer Key

1. B.
2. F.
3. B.
4. A.
5. C.
6. D.
7. C.
8. A.
9. B.
10. D.

Notes:

Q10. Answer options other than chloroplast are not organelles.

Chapter 7

The Cell Cycle

7.1 Cell Cycle: The Big Picture

You learned in Chapter 1 that one of the characteristics of life is reproduction, the ability of organisms to produce more of their own kind. Like all other functions, this fundamentally happens at the cellular level because cells are the basic building blocks of life. This means a cell reproduces itself. As we explored in Chapter 4, each new cell arises from an existing cell. All of us begin our life journey as a single cell, the fertilized egg. Subsequently, we grow into an organism with about a trillion cells. All these cells descend from that single cell through a repetitive process called cell division. Even after we grow up as adults, some cell reproduction continues through cell division in order to replace damaged and aged cells.

Cell division is only a part of the cell cycle, which is a series of events from the moment the cell forms from the division of its parent cell to the moment it divides itself into two daughter cells. So, the cycle is a journey from reproduction to reproduction. As illustrated in Figure 7.1, there are multiple types of cell division such as mitosis, meiosis, and binary fission. Mitosis is the cell division used by eukaryotic cells in which a cell divides into two daughter cells, leaving them capable of dividing themselves. This type of division is the basis for body growth, repair of damaged tissues, and replacement of tissues in animals as well as asexual

Cellular and Molecular Biology, by Paul Sanghera
Copyright © 2015 Infonential.

reproduction in some plants. Binary fission is the type of cell division that serves as asexual reproduction in prokaryotes and unicellular eukaryotes. In unicellular eukaryotes, mitosis is a part of binary fission. Meiosis is the type of cell division that functions as the basis for sexual reproduction in sexually reproducing organisms. It produces haploid cells, which are so named because they contain only half of the DNA of the parent cell and are incapable of dividing further.

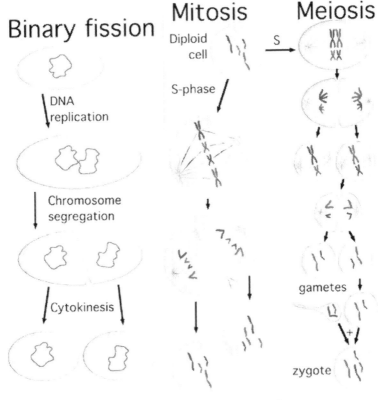

Figure 7.1 Three types of cell division.

The Cell Cycle

> **Caution!** Most cell division in both prokaryotes and eukaryotes results in cloned, that is, genetically identical daughter cells. Meiosis, the cell division that produces haploid cells called gametes (sperms and eggs), is an exception to this rule.

The steps of cell division are carefully regulated and controlled by a number of genes. When the regulation does not work correctly, health problems such as cancer can arise.

In this chapter, we will explore cell cycle, cell division, and regulation of the cell cycle.

7.2 The Cell Cycle: In Between Two Divisions

As illustrated in Figure 7.2, a cell cycle is a series of events in the life of a cell from one cell division where it comes into being to another cell division where it divides into two daughter cells which each begin their own cell cycle. The phase between two cell divisions is called *interphase*. The cell division phase called the mitotic phase

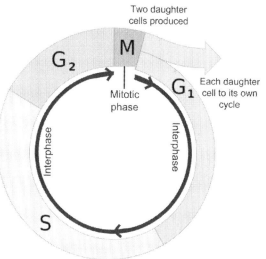

Figure 7.2 The cell cycle of a mitotically dividing cell.

lasts for only 10 percent of the cycle time in most cells and alternates with interphase, which accounts for the remaining

Cellular and Molecular Biology

90 percent of the cycle time. Interphase itself is divided into three subphases: G_1 phase, S phase, and G_2 phase.

G_1 **phase.** Also called the *first gap* or interval, this phase follows the mitotic phase. During this phase, the cell grows while performing some functions such as producing proteins and some cellular organelles. Organelles proliferate during this phase. Most of the cells of an organism remain in the G_1 phase doing jobs such as synthesizing carbohydrates, lipids, and proteins and supporting body functions such as a neuron (nerve cell) carrying a message (impulse). This phase is followed by the S phase. Cells headed for division enter the S phase.

S phase. During this phase, which is also called the synthesis phase, the cell synthesizes or duplicates its chromosomes, which involve DNA replication. The centrosome is also duplicated. The centrosome is a structure in the cytoplasm of animal cells that functions as a center to organize microtubules. It usually includes a pair of centrioles, which enable this functionality. A centriole is a barrel-shaped structure that helps the centrosome to organize microtubules and hence spindles. A spindle is a dynamic array of microtubules with two poles, which direct the motion of chromosomes during mitosis and meiosis. It is also called the bipolar spindle.

> **Fascinating Fact!** Our body has about 50 trillion diploid cells with 23 pairs of chromosomes in each cell. Each chromosome has one DNA molecule. It turns out that the total amount of DNA in each cell packed in these chromosomes is about 6 billion base pairs. Given that each base pair is about 0.34 nm long, if you stretch out the DNA from just one cell, it would extend about 2 meters. Therefore, the total amount of DNA in your body is about 100 trillion meters long (50 trillion cells multiplied by 2 meters per cell). This means that a typical human being has enough DNA that if stretched out can go from the Earth to the Sun and back more than 300 times.

This phase is followed by the G_2 phase.

G_2 phase. During this phase, which is also called the *second gap* or interval, the cell prepares for division. For example, it synthesizes proteins that later help in driving mitosis.

Cells of the same type have about the same length for the cell cycle duration, whereas cell cycle duration may differ from one cell type to another even within the same organism. For example, a red blood cell in your body divides every 12 hours, but adult neurons never divide. During the embryonic development of a sea urchin, the cell cycle is only 2 hours long.

> **Think About It!**
>
> Adult neurons never divide. Does it mean that all the neurons that make up the adult brain were produced during embryonic development and that the adult brain can never produce new neurons?
>
> Answer:
>
> Not necessarily. As a matter of fact, some parts of the brain have been found making new neurons from adult stem cells in the brain.

The G_2 phase is the last part of interphase, and it is followed by the mitotic phase during which mitotic cell division occurs.

7.3 Mitosis

Mitosis, the fundamental process for the development and growth of organisms, is used to make new body cells from

Cellular and Molecular Biology

existing cells. During this process, a cell splits into two genetically identical daughter cells. This process is also used for asexual reproduction by single parents to produce genetically identical offspring. Different phases of this process are illustrated in Figure 7.3 and described in the following:

Prophase. In the G_2 phase, chromosomes could not be seen individually because they had not yet condensed. In prophase, they condense into discrete coils and can be seen with a light microscope. These duplicated chromosomes are in pairs; each pair consists of two identical chromosomes called *sister chromatids* attached to each other at points called *centromeres*. The nucleoli disappear. Toward the end of this phase, the newly made two centrioles begin moving to the opposite sides of the nucleus, and the nuclear envelope begins to break down.

Prometaphase. The nuclear envelope breaks down. Microtubules extending from each of the two centrosomes constitute a bipolar spindle by penetrating the nuclear region. The spindle attaches to a region on the centromere of a sister chromatid called *kinetochore*, which is a structure of proteins. Chromosomes are condensed further.

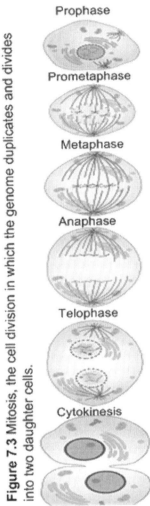

Figure 7.3 Mitosis, the cell division in which the genome duplicates and divides into two daughter cells.

The Cell Cycle

Metaphase. The two poles of the spindle are now at the opposite points of the cell. All of the duplicated chromosomes line up midway between the spindle poles on an imaginary plate called the metaphase plate, which is a plane that is equidistant from the two opposite poles of the spindle.

Anaphase. Each of the two chromatids of a duplicated chromosome separates and becomes a full chromosome. Directed by the microtubules, these two daughter chromosomes begin moving toward the opposite poles of the spindle. Microtubules push the poles farther apart, elongating the cell. Toward the end of this phase, the two sides of the cell have an equal number of chromosomes. Because the chromosomes were duplicated, each side has a complete genome of the organism to which the cell belongs.

Telophase. The fragments of the broken down nuclear envelope and membrane patches from other parts of the endomembrane system fuse together to form two nuclear envelopes around the two clusters of chromosomes, giving rise to the two daughter nuclei. Nucleoli in the two nuclei form. Chromosomes become less condensed. The mitotic division of one nucleus into two genetically identical daughter nuclei is complete. However, the division of the parent cell into two individual daughter cells is not yet complete. Toward the end of telophase, the division of the cytoplasm begins.

Cytokinesis. This is the process that divides the cytoplasm of a cell following the division of the cell nucleus, giving rise to two separate daughter cells. The process usually begins toward the end of mitosis, meiosis I, or meiosis II. How do the two daughter cells really separate? In an animal cell or in a cell of some algae, a cleavage furrow forms with the help of proteins such as actin and myosin, which in turn pinch the cell into two.

In plant cells, cytokinesis does not involve cleavage because the plant cells have a rigid wall around them in addition to plasma membrane. In this case, a new wall called a cell plate forms in the middle of the cell which is originated by the vesicles from the Golgi apparatus and brought to the middle of the cell by the microtubules. This cell wall grows, enlarges, and subsequently its surrounding membrane fuses with the plasma membrane around the parameter of the cell. This fusion and the forces involved in this fusion give rise to the two separate daughter cells.

> **Think About It!**
> In some organisms such as some algae and fungi, cells skip cytokinesis in the cell cycles that they repeatedly go through. What unique cell feature does this lead to?
>
> **Answer:**
> Cells skipping cytokinesis in cell cycles give rise to a large cell with multiple nuclei.

In mitosis, both diploid and haploid cells can divide. Each diploid cell divides into two daughter diploid cells, and each haploid cell divides into two daughter haploid cells. Mitosis occurs in nonsexual cells called somatic cells, which are all cells except gametes (sperms or eggs) or their precursors or parent cells.

However, only diploid cells and not haploid cells can undergo meiosis, which is explored in the next section.

7.4 Meiosis

Meiosis is the type of cell division that facilitates reproduction by creating egg and sperm cells, each of which contain only half of the chromosomes (and hence half of the genes) of the parent cell and therefore are called haploids. Many steps of meiosis are pretty much like those of mitosis. Just like in mitosis, the chromosomes in meiosis are replicated during interphase before meiosis begins. However, one replication is followed by two consecutive divisions in meiosis instead of the one division that occurs in mitosis. Figure 7.4 presents an overview of meiosis.

As illustrated in Figure 7.4, during meiosis, a diploid cell

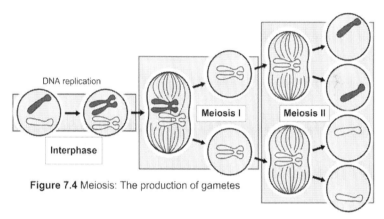

Figure 7.4 Meiosis: The production of gametes

replicates all of its contents, including its chromosomes. Subsequently, the cell splits to form two daughter cells in a stage called meiosis I. The daughter cells are haploid cells. In meiosis II, each of the two haploid daughter cells from meiosis I splits to form two daughter cells of meiosis II, which are also haploid. The steps in both meiosis I and meiosis II are very similar to those of mitosis, and they bear the same phase names with I and II appended to them for meiosis I and II respectively.

Here is a brief summary of the phases of meiosis I:

Prophase I. Condensation of chromosomes begins, and paternal and maternal homologs (each already duplicated) begin pairing along their length. Each pair of these two duplicated homologous chromosomes (four sister chromatids) is then held together. Recall that a homologous pair of chromosomes contains one chromosome from each parent. By now each of these chromosomes has been duplicated, and these two duplicated chromosomes are still connected to each other and are referred to as a *pair of homologous chromosomes*. Each member of this pair is called a homolog. This is not the case in prophase I, where each duplicated chromosome (two sister chromatids) is a separate entity.

Two homologs (parental and maternal) in a pair swap some DNA segments because they are so close to each other. This swapping is referred to as crossover or crossing. Recall that each homolog at this point is replicated and has two sister chromatids. Crossover occurs between two non-sister chromatids, paternal from one homolog and maternal from the other homolog.

Metaphase I. All of the pairs of homologous chromosomes line up on the metaphase plate just like the lineup that occurs during metaphase in mitosis. However, there is a very significant difference. In metaphase I, an individual in the lineup is a pair of homologous chromosomes composed of four sister chromatids, whereas in mitosis it is a single (duplicated) chromosome composed of two sister chromatids.

Anaphase I. Two homologs in each pair in the lineup separate, but the two sister chromatids in each homolog still remain attached to each other. Recall that one of these homologs is maternal and the other is paternal. Directed by the spindle, the maternal homolog begins moving toward one spindle pole, and the other one toward the opposite pole. Now note the difference: In mitosis, it is the single chromosome that travels toward a pole, whereas here it is the duplicated chromosome, an attached pair of sister chromatids, that travels to the pole.

Telophase I. Each of the two poles of a spindle now has either the paternal version of a chromosome or the maternal version but not both. However, each chromosome is still duplicated. In other words, each pole has enough material to make two haploid cells but not one diploid cell. To the contrary, in mitosis, each pole in telophase has enough material to make a diploid cell because it receives both the paternal version and the maternal version to form a homologous pair of chromosomes.

Also just like in mitosis, cytokinesis begins during telophase I, which splits the cytoplasm of the cell and gives rise to two haploid daughter cells.

There is no replication between meiosis I and meiosis II, which follows meiosis I. Here is a brief summary of the phases of meiosis II:

Prophase II. Just like the prophase of mitosis and prophase I, the bipolar spindle begins to form. The two sister chromatids in each homolog are still attached together. Homologs begin moving toward the metaphase plate in the middle of the cell.

Metaphase II. Just like in mitosis, the chromosomes (homologs) in metaphase II are lined up on the metaphase plate. Note that unlike in mitosis, the two sister chromatids in a homolog are not genetically identical. This is due to the crossover that happened during prophase I.

Anaphase II. Due to the forces of the spindle and the breakdown of the proteins holding the sister chromatids together, the chromatids separate. One chromatid, which is now a single unduplicated chromosome, moves toward one spindle pole, and the other one moves toward the opposite pole.

Telophase II. Now each of the two spindle poles has either the paternal version of a chromosome or the maternal version but not both. Also, each chromosome is unduplicated unlike the chromosomes in telophase I. Thus, each pole has enough material to make one haploid cell. A nuclear envelope forms around each of these two sets of chromosomes at the opposite poles, and cytokinesis splits the cell into two haploid daughter cells, which are not genetically identical. Note that in the whole process of meiosis, one diploid cell splits into four haploid cells. The daughter cells of meiosis II are eggs in a female and sperm in a male, and they are collectively called *gametes*. The replication in meiosis occurs once and cell splitting occurs twice, which reduces the chromosome number by half—from 46 to 23 in humans for example—to form sperm and egg cells. When the sperm and egg cells unite during fertilization, each contributes 23 chromosomes so that the resulting embryo will have the usual 46. As a result, two gametes (one from each parent) join to form a diploid cell called a zygote. This is where the life story of a new organism begins. This way, the process of meiosis (that is, meiosis I and meiosis II) ensures that organisms have the same number of chromosomes in each generation.

> **Caution!** In animals, only meiosis and not mitosis produces gametes (sperms and eggs). However, in plants, both mitosis and meiosis can produce gametes.

While cells are dividing, meiosis allows genetic variation through a process called DNA shuffling.

7.5 Contribution of Meiosis to Evolution

Genetic variations are the raw material on which natural selection, a mechanism of evolution, operates. Meiosis has contributed to evolution by introducing genetic variations through crossover, which was discussed in the previous section, and independent assortment, which is discussed in the following.

Independent assortment refers to the fact that during meiosis, genes on each pair of homologous chromosomes get sorted for the two gametes independent of how genes on any other pair of homologous chromosomes are sorted out. Recall that in metaphase I, all of the pairs of homologous chromosomes line up on the metaphase plate. One of these homologs is a duplicated paternal chromosome, and the other one is the duplicated maternal chromosome. Both the paternal homolog and the maternal homolog have equal chances to be closer to a given spindle pole. Therefore, a specific daughter cell of meiosis I has a 50% (0.5) probability that it will receive the maternal homolog and a 50% probability to receive the paternal homolog. Because each pair of homologs is positioned on the metaphase plate independent of all other pairs, the sorting of homologs in a daughter cell for one pair occurs independently of the sorting in all other pairs. This is what is called independent assortment.

Problem 7.1 Calculate the total number of possible combinations of genes in a given human sperm or egg. When the sperm or egg forms, one of these possibilities (combinations) materializes.

Solution:

Total number of combinations for one homologous pair = 2

A given gamete (egg or sperm) will get either a maternal chromosome or a paternal chromosome.

This sorting for each pair of homologs is independent of sorting in all other pairs.

Therefore, the total number of combinations = 2^{23} = 8.39 million.

As shown in Problem 7.1, each human sperm or egg contains one out of about 8.4 possible sets (combinations) of genes. This randomness, due to independent assortment, contributes to genetic variations from one generation to the next. The crossover that happens during prophase I also contributes to genetic variations. At a population level, this random variation contributes to evolution.

> **Caution!** Even though independent assortment and crossover in meiosis add to genetic diversity, they are not the original source of genetic diversity. Gene mutations are the original source of genetic variations.

7.6 Regulation of the Eukaryotic Cell Cycle

The regulation of the cell cycle, including cell division, is critical to the growth and reproduction of organisms. For example, the human body itself develops out of a single cell, the zygote, through cell division. Even when you become an adult, millions of cells in the different parts of your body such as bone marrow, blood, gut lining, liver, and skin are multiplying though cell division to replace dead or worn-out cells every second. Cell division does not occur randomly; it is a well-regulated process.

> **Growth Factors.** The signals participating in controlling the cell cycle may originate inside the cell or outside the cell. An example of external signal molecules is the protein called growth factor, which is released by certain cells to stimulate other cells to divide.

The eukaryotic cell cycle is regulated by the molecular system constituted by the signaling molecules present in the cell cytoplasm. Some proteins such as cyclins and cyclin-dependent kinases (Cdks) function as a cell cycle clock to pace or schedule the events of the cell cycle. This clock supports built-in checkpoints such as G_1, G_2, and M checkpoints at G_1, G_2, and M phases of the cycle. The cell cycle halts at a checkpoint until a go-ahead signal is received. The protein kinases constitute go or stop signals for the checkpoints by activating or inactivating specific proteins. After receiving a go-ahead signal at a G_1 checkpoint, a cell in most cases completes the whole cycle: G_1, S, G_2, and M phases. If the cell does not receive the go-ahead signal at the G_1 checkpoint, it falls into a state called the G_0 phase in which it stops dividing. For example, most cells in an adult human body are in the G_0 phase. As mentioned earlier, many of these cells are called back for healing and repair work. In mammalian cells, the G_1 checkpoint is also referred to as the restriction point.

> **Fascinating Facts!** Your body heals itself from a cut or wound, which disappears in a week or two. We regenerate our skin every week and our liver about every year. Each type of tissue has its own turnover time depending on various factors, including the workload endured by its cells. Nevertheless, the entire human skeleton is replaced in about 7 to 10 years. And all this happens spontaneously through cell division; no surgeries are involved.

Cellular mechanisms at a given checkpoint determine if the cell is prepared to continue with the cycle. For example, at the G_2 checkpoint, it is determined whether the cell has enough specific kinases called *maturation promoting factor* (MPF) to direct mitotic events in the next phase, the M phase. Similarly, at the M checkpoint it is determined if all the chromosomes are aligned properly at the metaphase plate. Only after confirming this alignment does the cell cycle proceed to separate the sister chromatids and allow them to continue their journey to the spindle poles.

As you already know, the production of proteins that control the cell cycle is an expression of certain genes.

Cellular and Molecular Biology

Certain gene mutations can cause the production of faulty proteins, which will result in the failure of the cell cycle control system. Such failure can cause uncontrolled cell growth, which may result in benign (noncancerous) moles or cancerous tumors. Benign tumors are localized; that is, they remain confined to their original site and hence can be easily removed with surgery. To the contrary, cancerous tumors called malignant tumors spread to the neighboring tissues and can impair one or more organs.

7.7 In a Nutshell

- The cell cycle is a sequence of events in the life of a cell from the moment it comes into being from the division of its parent cell to the moment of its own division into two daughter cells.
- The cell cycle is divided into two phases: interphase and mitotic phase.
- During interphase, the cell grows, replicates all its chromosomes, and performs some cellular functions such as metabolic functions.
- Meiosis I divides a diploid cell to two haploid daughter cells, whereas meiosis II divides a haploid cell to two haploid daughter cells.
- Meiosis contributes to evolution by generating genetic variations.

7.8 Review Questions

1. Which of the following constitutes a complete cell cycle?
    A. interphase and mitotic phase

B. interphase, prophase, prometaphase, metaphase, anaphase, telophase, and cytokinesis
   C. G_1 phase, S phase, G_2 phase, interphase
   D. interphase and meiosis
   E. A., B., and C.

2. Interphase and mitotic phase ____.
   A. constitute one turn of a cell cycle.
   B. constitute the division of a cell.
   C. take the same time in all cells of a given organism.
   D. constitute a cell cycle in all kinds of organisms.

3. Which of the following is a true statement?
   A. Only diploid cells can use mitosis.
   B. Cytokinesis is only used in animals but not in plants.
   C. Cleavage furrow is the process used by both plants and animals to divide a parent cell into two daughter cells.
   D. Only diploid cells can undergo meiosis.

4. Which of the following statements is not correct?
   A. Meiosis I divides a diploid cell into two haploid cells.
   B. The two sister chromatids in a chromosome lined up on the metaphase plate are genetically identical in both metaphase of mitosis and metaphase II of meiosis II.

Cellular and Molecular Biology

   C. Meiosis II divides a haploid cell into two haploid cells.
   D. In addition to other uses, mitosis is also used for asexual reproduction.

5. Consider an embryo developed from six cell divisions of a zygote. How many cells are in the embryo?
   A. 6
   B. 12
   C. 64
   D. 128

6. In the cell cycle, which of the following does not happen before the mitotic phase begins?
   A. Chromosomes are duplicated.
   B. Cellular organelles proliferate.
   C. A bipolar spindle forms.
   D. Carbohydrates, lipids, proteins, and nucleic acids are synthesized.

7. Two copies of a chromosome duplicated during interphase which are connected to each other and line up on the metaphase plate are called ____.
   A. homologs
   B. sister chromatids
   C. centrioles

D. gametes

8. During which stage of cell division does crossover occur?
   A. prophase II
   B. metaphase I
   C. anaphase I
   D. prophase I

9. In which phase of cell division does the bipolar spindle disintegrate?
   A. telophase
   B. cytokinesis
   C. interphase
   D. anaphase

10. A normal human cell has how many DNA molecules in the G_2 phase?
    A. 23
    B. 46
    C. 92
    D. There is no way to determine this information.

11. A cell usually completes the cell cycle and divides if it gets a go-ahead signal at the checkpoint in which phase?
    A. G_1

Cellular and Molecular Biology

B. S
C. M
D. G_2

12. Which of the following incorrectly matches the phase of the cell cycle with its function?

    A. G_1: organelles proliferate
    B. M: DNA replication
    C. G_2: synthesizes proteins that help driving mitosis
    D. S: duplicates centrosome

7.9 Answer Key

1. E.
2. A.
3. D.
4. B.
5. C.
6. C.
7. B.
8. D.
9. A.
10. C.
11. A.
12. B.

Notes:

Q4. The two sister chromatids in a chromosome (homolog) in metaphase II are not genetically identical. This is due to crossover that happened during prophase I.

Q5. An embryo develops through mitotic division. Starting from one cell (zygote), five divisions will create $2^5 = 64$ cells.

Q10. The human genome has 23 pairs of chromosomes, that is, 46 chromosomes; each chromosome has one DNA molecule; and chromosomes are duplicated in the S phase preceding the G_2 phase.

Cellular and Molecular Biology

Chapter 8

Classical Genetics

8.1 Emergence of Mendelian Genetics: The Big Picture

As discussed in Chapter 7, meiosis halves the chromosome number when making sperms and eggs, the gametes, which are therefore called haploid cells. The offspring receives half of its chromosomes from each of its two parents in the form of a zygote, a fusion of sperm and egg during fertilization. Some characteristics (or traits) are inherited from parents through the genes in DNA molecules, which are packed in these chromosomes. Now we know it, but when Darwin and Wallace presented their theory of evolution in 1858, scientists were still struggling to figure out how inheritance works.

Figure 8.1 Mendel's research garden.

The prevalent view of inheritance in those times was the so-called blending hypothesis. According to this hypothesis,

Molecular and Cellular Biology, by Paul Sanghera
Copyright ©2015 Infonential.

the hereditary material from both parents blends together like a fluid at fertilization in a uniform fashion. Think of red wine mixing with grape juice or vodka mixing with orange juice. This means that the offspring should acquire the characteristics intermediate between their parents. However, this hypothesis has a very vivid prediction: After many generations, a freely (randomly) mating population will give rise to a population of individuals that have very uniform characteristics; that is, all individuals look alike, and variations tend to cease. But the evidence against this prediction and therefore against the blending hypothesis in general are all around. Even the children of the same parents do not look exactly alike unless they are identical twins. The blending hypothesis also fails to explain many other facts of inheritance such as the reappearance of some traits after skipping generations. For example, a child may not get the attached earlobes of her mother, but the grandchild may. Furthermore, breeding experiments have also contradicted this prediction. This problem occupied Darwin because his theory of evolution was based on the fact that traits vary among individuals in a population. The trait variants that help individuals survive and reproduce in a given environment better than alternative trait variants become more common among the population over generations. However, what is the origin of these trait variants, and how exactly are they inherited? These were the questions that Darwin and other scientists were struggling with, and the blending inheritance hypothesis did not provide much help.

Even though his work remained unknown to Darwin and most other scientists during his lifetime, an Austrian monk named Gregor Mendel figured out the answers to the burning inheritance questions of their time. While working on the inheritance of trait variations in plants, three botanists—Carl Correns, Erik von Tschermak, and Hugo de Vries – rediscovered Mendel's work in 1900. Once brought to light, Gregor Mendel's now famous breeding experiments with peas eliminated the confusion and dilemma created by the idea of blending inheritance. The actual experiments were performed in late 1850s and early 1860s in his monastery's two hectare experimental research garden (Figure 8.1), which was originally planted by the abbot Napp in 1830. From 1856 to 1863, Mendel cultivated and tested about 29,000 pea plants. Mendel's garden displayed obvious morphological differences between different types of peas: smooth versus wrinkled seeds, short versus tall stems, purple versus white flowers, and so on. The main results from these experiments directly challenged the blending hypothesis. Due to his scientific contribution, Mendel is now remembered as the father of modern genetics.

From the results of his experiments, Mendel developed his model of inheritance or genetics now known as Mendelian genetics, which we will explore in this chapter.

8.2 Mendel's Model of Genetics

Around 1860, Gregor Mendel experimented with pea plants in his monastery garden to learn about heredity. For example, in one of his experiments, which is illustrated in Figure 8.2, Mendel crossed true breeding purple-flowered pea plants with true breeding white-flowered pea plants. The true-breed generation is called the parent (P) generation. In the first offspring generation (F_1), all pea plants turned out to be purple-flowered. When Mendel self-pollinated F_1 plants or crossed F_1 plants with other F_1 plants, the offspring (F_2 generation) turned out to be about 75% purple-flowered and 25% white-flowered. To explain the results of his experiments such as the 3:1 ratio in the F_2 generation which he observed for many traits, Mendel developed his model which has the following main elements:

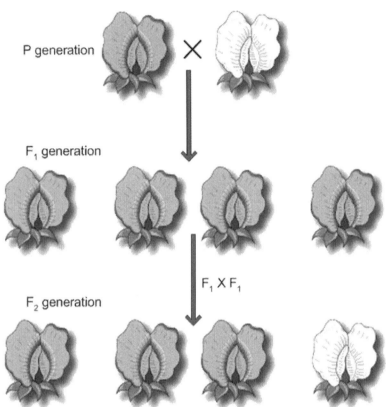

Figure 8.2 Crossing of pea plants: an example of Mendel's experiments

Traits linked to genes, variations linked to alleles. Inherited characteristics or traits of organisms are linked to genes inherited from the organism's parents. Variations in inherited traits are due to alternate versions of genes that may be inherited. For example, the white and purple colors of pea plant flowers correspond to two versions of the color gene. The alternate versions of a gene are called alleles.

Classical Genetics

Both parents contribute. For each trait, an organism inherits two alleles: one from each parent. For example, during meiosis, the allele for a gene inherited from the male parent will get into the sperm, and the allele for the same gene inherited from the female parent will come through the egg. As you know from Chapter 7, this choice is made randomly, that is, according to the probability laws. Then why did all the F_1 generation peas turned out to be purple-flowered? That has to do with dominance and recession. Believe it or not, in olden times, it was not obvious to scientists that both parents contribute genes to genetic inheritance; females were considered to be noncontributors by many.

Dominance and recession. If the two inherited alleles are different from each other, then one of them is dominant, and the other one is recessive. The dominant one determines the trait of the organism. This explains the fact that all the pea plants in F_1 generation are purple-flowered: the allele for the purple flower is dominant.

1. **Law of segregation.** During gamete formation, two alleles corresponding to a heritable trait separate (segregate) as part of chromosomes and enter different gametes. This happens during meiosis, which is discussed in Chapter 7. This means that an egg gets only one of the two alleles corresponding to a trait. The same is true for the sperm.

2. **Law of independent assortment.** During gamete formation, each pair of alleles segregates independently of other pairs of alleles. This law is discussed in the form of chromosomes in Chapter 7. Alleles are in DNA molecules, which in turn are in chromosomes. Independent assortment basically comes into action in metaphase I of meiosis when homologous chromosome pairs align along the metaphase plate. The alignment (which of the homologs face which pole) of each chromosome pair occurs

independently of the alignment of any other pair. This is the source of the independent assortment of alleles.

Thus, genetic characteristics are inherited in terms of alleles.

8.3 Inheritance of Characteristics in Terms of Alleles

In the framework of Mendelian genetics, inheritance of traits can be understood and predicted in terms of alleles. In order to be able to explore this aspect, let us first explore some genetic terminology related to crossing and the resulting offspring by using Figure 8.3.

Phenotype. A phenotype is an observable physical and physiological characteristic (trait) of an organism determined by a set of genes or alleles (genetic makeup) of the organism. Some examples of phenotypes are hair color, eye color, and height of an organism. In the figure, the colors (pink and white) are phenotypes.

Genotype. A genotype is a set of alleles of an organism contributing to determining a phenotype. In the figure, Pp, PP, and pp are genotypes.

Alleles. These are different versions of the same gene that contribute to different phenotypic features such as blue eyes and dark eyes. Note that alleles contribute to the diversity of phenotypic features. In the figure, P and p are alleles for the color gene: P represents purple, and p represents white.

Homozygous. An organism having two identical alleles for the same gene corresponding to a trait is said to be homozygous for that gene or trait. For example, the white-flowered pea plant in the bottom right corner of the figure is homozygous (pp) for the recessive allele p, and the purple-flowered pea plant in the top left column is homozygous (PP) for the dominant allele P. An organism homozygous for a gene is called a *true breed* for that gene.

Heterozygous. An organism having two different alleles for the same gene corresponding to a trait is said to be heterozygous for that gene or trait. For example, the purple-flowered pea plants in the F_1 generation in the figure are both heterozygous (Pp) for the purple flower color gene.

Monohybrid/dihybrid. A monohybrid is a cross to study the inheritance of one trait such as color, whereas a dihybrid is a cross to study the simultaneous inheritance of two traits such as color and height.

The genotypes (and therefore phenotypes) of the offspring from the genetic crosses of two organisms with known genotypes can be predicted with some probability constructing a matrix called a Punnett square. Figure 8.3 is an example of a Punnett square that predicts the genotypes of the offspring from the self-crossing of the F_1 generation of pea plants discussed in the previous section. In the following, we use this figure as an example to describe the general steps to build a Punnett square for a crossing.

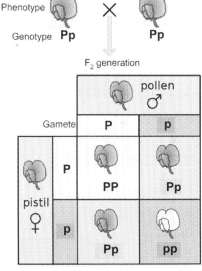

Figure 8.3. Punnett square to predict F_2 generation from the self-crossing of F_1 generation.

1. Write down the genotype of both organisms that are being crossed such as Pp and Pp.

Molecular and Cellular Biology

2. Split the genotype of each organism to alleles: P, p and P, p in our example.

3. Place the alleles of one organism (say female) along the rows of matrix and the corresponding alleles of the other organism (say male) along the columns of the matrix as shown in Figure 8.3.

4. Fill in the cells of the table (matrix) with all the combinations of the alleles along the rows with the alleles along the columns.

Each cell contains one possible genotype outcome of the crossing between the two organisms. The Punnett square lists all the possible outcomes, and the probability of a specific outcome can be calculated from the square. In Figure 8.3, because both PP and pp occupy one out of four blocks, and Pp occupy two out of four blocks, the probability that the offspring of Pp x Pp crossing will have PP genotype is 0.25, the probability that the offspring will have genotype pp is also 0.25, whereas the probability for the Pp genotype is 0.50.

We discussed the law of independent assortment very briefly in the previous section. Let us explore it further.

8.4 Mendel's Law of Independent Assortment

The law of independent assortment discussed in terms of chromosomes and gametes in Chapter 7 was originally derived by Mendel from his experiments when he did not know chromosomes or even genes the way we know them today. According to independent assortment, during gamete formation, each pair of alleles segregates independently of each other pair of alleles. This means that the alleles for the gametes are selected randomly from all possible combinations with the constraint that a gamete has one allele for each gene. We elaborate upon independent assortment in the framework of Mendelian genetics by using the example illustrated in Figure 8.4. In this example, we consider the simultaneous inheritance of two traits (dihybrid crossing): coat color (B for brown which is dominant and b for white which is recessive), and tail length (S for short which is dominant and s for long which is recessive). Even though Mendel performed his experiments on peas, we are considering cats in this example just to show you that these laws are general because the genetic code is general. Let us explore this example in Figure 8.4:

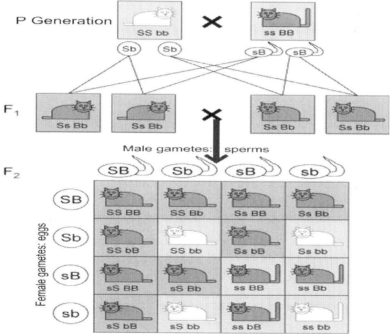

Figure 8.4 Mendel's law of independent assortment illustrated for dihybrid crossing.

Parents: P generation. Parents in this example are homozygous in both traits: SS and bb for short white cat and ss and BB for long brown cat.

F1 generation. As shown in Figure 8.4, the F_1 generation from this cross is heterozygous in both traits. The P generation can make only one type of gamete for each parent: Sb for short white and sB long brown. Therefore all the offspring, that is, the F_1 generation from these gametes will be of genotype SsBb. Individuals from this generation are crossed with each other to produce the F_2 generation.

F2 generation. The F_2 generation inherited two types of gametes (pairs of alleles): Sb and sB. If independent assortment is in action, alleles in these gametes can make any of all possible combinations during their meiosis. Therefore, F_1 self-crossing will produce four types of gametes in equal quantities: Sb, SB, sb, and sB. Now in random fertilization, four types of sperms will combine with four types of eggs, giving rise to 16 (4x4) types of possible combinations, that is, zygotes or organisms with equal probabilities, as shown in the figure. This results in producing a phenotypic ratio of 9:3:3:1 where 9 are brown short (dominant/dominant), 3 are brown long (dominant/recessive), 3 are white short (recessive/dominant), and 1 is white long (recessive/recessive).

Now that we know more about chromosomes, we also know that the law of independent assortment applies only to allele pairs that exist on chromosomes that are not homologous to each other or are very far apart on the same homologous pair. When they are far apart on the same homologous pair of chromosomes, they appear as if they are on different chromosome pairs due to crossover. It turns out that the pea traits that Mendel selected for his experiments meet this requirement: These traits are represented by allele pairs that exist on different homologous pairs of chromosomes or on the same homologous pair but far apart.

Classical Genetics

> **Problem 8.1** Imagine a world where independent assortment is not the law of nature. Instead, dependent assortment is in action. What will the F_2 generation in the example illustrated in Figure 8.4 look like in that world? Draw the Punnett square for this situation.
>
> **Solution:**
>
> As shown in Figure 8.4, only two possible types of gametes to produce the F_1 generation are Sb and sB. This remains true under both independent and dependent assortment. The difference kicks in for the production of gametes in the F_1 generation to produce the F_2 generation. This production under independent assortment is predicted in Figure 8.4, and
>
> | | Male gametes (sperms) | |
> | --- | --- | --- |
> | Female gametes (eggs) | Sb | sB |
> | Sb | SSbb
Short white | SsBb
Short brown |
> | sB | SsBb
Short brown | ssBB
long brown |
>
> Punnett square to predict F_2 generation in Figure 8.4 based on dependent assortment.
>
> the production under dependent assortment is predicted in the figure in this box. Under dependent assortment, the F_1 generation will transmit the alleles in the same combinations in which they inherited them from their parent generation. This means the gametes for the F_2 generation remain of two types: Sb and sB. This will produce the F_2 generation predicted in the figure in this box. As shown in the figure, it will result in the production of short brown, short white, and long brown individuals in the ratio of 2:1:1 and no long white type.

And now we know that the inheritance patterns in general can be more complex.

8.5 Extending Mendel's Model to More Complex Situations

In simple Mendelian genetics discussed so far in this chapter, each heritable trait is determined by one gene for which there are only two alleles, one completely dominant and the other completely recessive; that is, the dominant allele completely masks the effect of the recessive allele. These two alleles reside on two homologous chromosomes, one on each, which were inherited from each parent. However, there is no role for the environment in Mendelian genetics in deriving phenotypes from genotypes. This simplest version of the model works for some traits such as those that Mendel studied, but it does not work for all traits.

Complex inheritance patterns may arise from both single genes and a group of genes.

8.5.1 Complex inheritance patterns determined by a single gene

Even though he could not understand and explain complex inheritance patterns, Mendel was already ahead of his time. The principles of segregation and independent assortment that Mendel derived from his work can be applied in general forms to complex patterns. After discovering his work at the beginning of the twentieth century, scientists found that Mendelian principles can easily be extended not only to more complex patterns but also to other organisms in addition to plants.

Let us explore the extension of Mendelian genetics for traits determined by single genes. According to basic Mendelian genetics, one of the alleles is completely dominant and the other one completely recessive, and the trait is determined by the dominant alleles when both are present in a gene. For some traits, this pattern becomes more complex in following ways:

Incomplete dominance. This is the inheritance pattern in which the phenotype of a heterozygote is an intermediate of the phenotypes of two homozygotes for each of the two alleles. For example, true red (RR) and true white (WW) flowers crossed with each other produce pink flowers (RW) in the F_1 generation. In this case, R refers to the allele for red color, and W stands for the allele for white color. You can verify this by drawing a Punnett square that shows that the self-crossing of the F_1 generation (RW x RW) will produce the F_2 generation with a 50% probability for being pink, a 25% probability for being red, and a 25% probability for being white.

Codominance. Codominance is the inheritance pattern in which a heterozygote contains the phenotypes corresponding to both alleles. For example, in the AB blood group, both alleles (i^A and i^B) are equally dominant in making the blood.

Multiple alleles. Most genes have more than two possible alleles. For example, the gene that determines the blood type (phenotype) has three alleles: i^A and i^B, and i. This means that each of the two alleles of a gene may be selected from these three types. This gives rise to six (2x3) possible combinations (genotypes) resulting in four possible blood types (phenotypes): blood type A ($i^A i^A$ or i i^A), blood type B ($i^B i^B$ or i i^B), blood type AB ($i^B i^B$), and blood type O (ii). As shown in the solution of Problem 8.2, the Punnett square for single genes with more than two alleles can be drawn exactly like the one for single genes with two alleles.

Pleiotropy. This is the ability of a single gene to have multiple phenotypic effects. For example, the gene in a pea plant that determines the color of the pea flower also affects the color of the outer coating of the seed.

Molecular and Cellular Biology

> **Problem 8.2** Consider the cross between a male parent with blood type AB and a female parent who is heterozygous for blood type B. Draw the Punnett square for this cross and predict the probability of blood types for the offspring.
>
> **Solution:**
>
> Genotype of male parent: $i^A i^B$
>
> Genotype of female parent: $i^B i$
>
>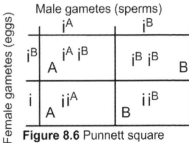
>
> **Figure 8.6** Punnett square showing all possible allele combinations for the offspring that result from an AB X BO cross.
>
> From the Punnett square drawn in this box:
>
> Probability for blood type A: 50% = 0.50
>
> Probability for blood type B: 50 % = 0.50

In this section, we explored some situations of complex inheritance when a trait is determined by a single gene. Another type of complex inheritance arises when multiple genes, as opposed to a single gene, determine a trait.

8.5.2 Complex inheritance patterns determined by multiple genes

Here are two of the ways multiple genes can determine a trait. One of them can interfere (or mask) the phenotypic effect of the other, or both of them contribute to a trait such as creating a continuous variation of the trait (phenotype). In the following, we discuss these two cases of complex inheritance patterns arising from the situations where a trait is determined by multiple genes.

Epistasis. This is a type of gene interaction in which one gene alters the phenotypic effect of another gene that was independently inherited. For example, if an individual is a homozygote for a dark color and also a homozygote for a recessive pigment with white color, the individual will not receive the dark color. It is a condition called *albinism*.

> **Caution!** Note that epistasis is not the same inheritance pattern as dominance. In epistasis, an allele (which may be mutated) in one gene masks or affects the expression of another gene, whereas in epistasis, one allele of a gene masks the expression of another allele of the same gene.

Polygenic inheritance. This is the inheritance pattern in which multiple genes contribute to a single phenotypic trait; it is the converse of pleiotropy. The superposition of effects from multiple genes determines the phenotype. In this case, the phenotype is usually a continuum such as skin color or height rather than an *either-or* trait such as round pea versus wrinkled pea.

As an example, skin color is determined by more than two inherited genes, which produce a continuous variation of skin color across individuals of a population.

8.5.3 Environmental factors

Even though inherited traits are based on genotypes, environment does play a role in translating genotypes to phenotypes. Depending on the traits, the effect of environment varies widely. For some traits such as the blood type, environment has a minimum role once the genotype is determined, whereas for other traits such as skin color, height, and intelligence, environment may play a significant role. For example, in the absence of proper nutrition, a child with tall genes may not grow as tall, and you can alter the color of your skin with sun tanning. A genetically smart child may fail an exam without enough preparation. Some plants have been found to have a flower color that is different from what is genetically expected. This may happen, for example, due to the acidity (pH) and other content of the soil from which they draw their nutrition.

In general, the role of environment in the relationship between genotypes and phenotypes can be expressed by the following equation:

$$\text{Genotype} + \text{Environment} \Longrightarrow \text{Phenotype}$$

8.6 In a Nutshell

From the study of pea plants, Mendel derived laws of inheritance such as the law of segregation and the law of independent assortment. These laws have a wider scope beyond just peas or plants due to the universality of the genetic code. Even though these laws were derived from simple inheritance patterns, they can be easily extended to more complex inheritance patterns.

8.7 Review Questions

1. A cross between a female with black hair and brown eyes and a male with blond hair and blue eyes ____.

    A. cannot be handled in the framework of Mendelian genetics.

    B. is called a monohybrid.

    C. is called a dihybrid.

D. is called a quadruple hybrid.

2. The independent assortment of alleles is enforced in ____.
   A. the S phase of the cell cycle
   B. the M phase of the cell cycle
   C. meiosis
   D. the G_2 phase of the cell cycle

3. How many types of gametes can be produced by an individual corresponding to the genotype AABb?
   A. 2
   B. 3
   C. 4
   D. 8

4. An organism is called a heterozygote for a trait when ____ in the gene for that trait.
   A. it has a pair of nonidentical alleles
   B. it has a pair of identical alleles
   C. both of the alleles are recessive
   D. it has a mutation

5. The blood type AB is an example of ____.
   A. dihybrid

Molecular and Cellular Biology

    B. codominance

    C. incomplete dominance

    D. epistasis

6. The inheritance pattern in which one gene alters the phenotypic effect of another gene, is called ____.

    A. polygenic inheritance

    B. epistasis

    C. codominance

    D. pleiotropy

    E. dominance

7. Independent assortment of alleles fundamentally happens during ____.

    A. metaphase

    B. anaphase of mitosis

    C. metaphase II

    D. metaphase I

8. Which of the following inheritance patterns does not involve more than two genes in determining a trait?

    A. dihybrid

    B. epistasis

    C. polygenic

    D. albinism

9. The plants of the same species with the same genotype for flower color are planted in a garden with two lots: A and B. In lot A, the flowers are green, and in lot B, the flowers are blue. This anomaly can be explained by ____.

   A. gene mutations

   B. codominance

   C. polygenic inheritance

   D. environmental factors

10. Crossover happens ____.

    A. between sister chromatids of a homolog

    B. between two non-sister chromatids of homologs

    C. during metaphase I

    D. during anaphase I

Molecular and Cellular Biology

8.8 Answer Key

1. C.
2. C.
3. A.
4. A.
5. B.
6. B.
7. D.
8. A.
9. D.
10. B.

Notes:

Q3. Only two types of gametes can be produced in this case: AB and Ab.

Q7. Independent assortment does not happen in mitosis; it happens during metaphase I of meiosis when each homologous pair of chromosomes aligns on the metaphase plate with equal probability for each chromatid of the pair to face each of the two spindle poles. Each pair does this independently of all other pairs.

Q10. Crossover happens between non-sister chromatids of two homologs during prophase I.

Chapter 9

Molecular Genetics

9.1 Molecular Genetics: The Big Picture

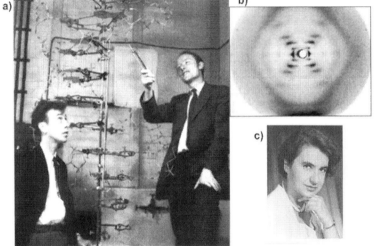

Figure 9.1 James Watson (your left) and Francis Crick (a) discovered the structure of the DNA molecule based on the X-ray diffraction pattern (b) taken by Rosalind Franklin (c). Courtesy of the U.S. National Institute of Health.

Francis Crick, James D. Watson, and Maurice Wilkins (a colleague of Rosalind Franklin) were jointly awarded the 1962 Nobel Prize for Physiology or Medicine for their discoveries of the structure of DNA. Franklin did not share the Nobel Prize as she had died in 1958 at the age of 37, and

Molecular and Cellular Biology, by Paul Sanghera
Copyright © 2015 Infonential.

Molecular and Cellular Biology

the Nobel Prize rules state that each recipient must be alive to receive the prize.

According to a legend, Francis Crick, originally a physicist and then a biologist as well, and James Watson, a biologist, walked into the Eagle Pub in Cambridge, England, on February 28, 1953, and Crick announced, "We have found the secret of life." What they had really figured out was the now famous double helix structure of the DNA molecule. This discovery of DNA structure started a revolution in biology by accelerating the already existing fields of molecular biology and biotechnology like a catalyst. Nevertheless, there was still a gigantic gap between determining the structure of DNA and understanding the exact role of DNA. It took the next quarter century of Crick's life along with other scientists to develop the ideas that now constitute the basics of molecular biology: the genetic code, messenger ribonucleic acid (mRNA), and the translation of mRNA into proteins. More than half a century after that event in the Eagle Pub, the DNA molecule is still the biggest celebrity in biology.

In Chapter 3, we explored four molecules of life: proteins, carbohydrates, lipids, and nucleic acids (DNA and RNA). Nucleic acids are unique in many respects. They have the built-in capability to replicate themselves, and they have genetic material (genes) that is transferred from one generation to the next. In other words, genetic traits are transferred from generation to generation through DNA, and this is possible due to its precise replication capability. Proteins are synthesized according to the genetic instructions in DNA, and the whole organism is developed according to the blueprint in the DNA code.

The study of genes including their effects at the molecular level is called molecular genetics. In this chapter, we explore molecular genetics from the anatomy and physiology of the DNA molecule to synthesizing proteins by expressing genes in DNA and regulating gene expression.

9.2 DNA Structure

We discussed DNA structure in Chapter 3. Here we review some important aspects of its structure.

Elements of DNA structure. DNA is genetic material, the molecule that contains hereditary information encoded in it in the form of a sequence of building blocks called nucleotides. The elements of DNA structure are illustrated in Figure 9.2. A nucleotide, the building block of nucleic acid, is generally composed of a 5-carbon (pentose) sugar covalently bonded to a nitrogenous base and a phosphate group. In a DNA molecule, the sugar is deoxyribose, whereas in an RNA molecule, the sugar is ribose.

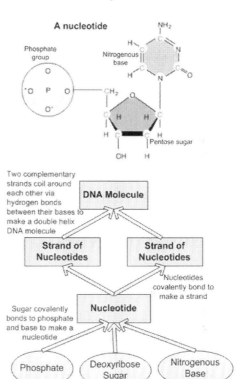

Figure 9.2 Composition of a DNA molecule.

Morphology. As illustrated in Figure 3.8 of Chapter 3, a DNA molecule is a double helix with sides (or backbones) made of sugar and phosphate.

Molecular and Cellular Biology

Strands bonded together. The nitrogenous bases of one strand are linked to the nitrogenous bases of the other strand with bonds called hydrogen bonds, which are discussed in Chapter 2. Base A always bonds with T, and G always bonds with C. This is why the number of As is equal to the number of Ts, and the number of Gs is equal to the number of Cs in a DNA molecule of any organism.

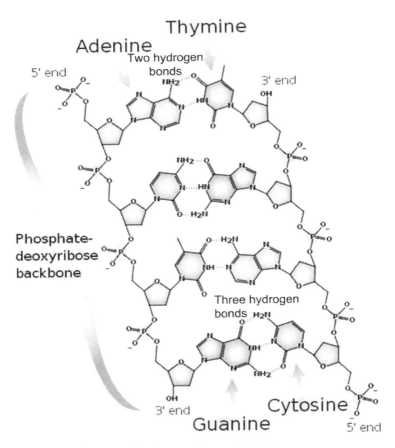

Figure 9.3 Chemical structure of DNA

Complementary strands. As illustrated in Figure 9.3, the two strands are antiparallel, that is, complementary to each other: one runs in the 5'-3' direction and the other in the 3'-5' direction. The numbers 3' and 5' come from the position of carbon in the deoxyribose sugar.

Base pairs. As illustrated in Figure 9.3, the two bases, one from each strand, bonded together are collectively called a base pair. Bases A and T are bonded together with double hydrogen bonds, whereas G and C are bonded to each other with triple hydrogen bonds. As illustrated in Figure 3.9 of Chapter 3, bases A and G consist of two rings and are called purines, whereas C and T consist of only one ring each and are called pyrimidines. The fact that A and G (two rings each) always bond with T and C (one ring each) respectively is also responsible for the uniform diameter of about 2.5 nm for the double helix.

> **Fascinating Fact!** The only difference in the basic composition of RNA and DNA is that RNA has ribose sugar instead of deoxyribose sugar in DNA, and RNA has a uracil base (U) instead of a thymine base (T). The only difference between ribose and deoxyribose is of one atoms: deoxyribose is missing an oxygen atom that ribose has, and hence it has the name deoxyribose.
>
>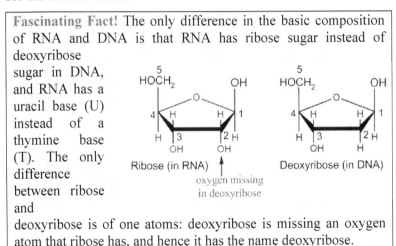

Molecular and Cellular Biology

The consecutive base pairs are about 0.34 nm apart, whereas one full turn of the helix contains 10 base pairs and is therefore 34 nm in length.

As you already know by now, the DNA molecule comes in a chromosome. How does this happen? Let us explore.

9.3 DNA Packaging Into a Chromosome

As DNA is created through replication, it is packaged tightly with proteins. The complex of DNA and proteins is generally called *chromatin*. As illustrated in Figure 9.4 and described in the following, eukaryotic DNA displays four levels of packaging: nucleosome, chromatin fiber, looped domain, and chromosome.

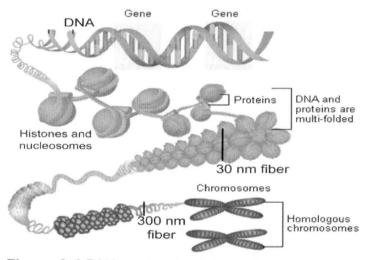

Figure 9.4 DNA packaging into a chromosome

Nucleosome (10 nm structure). The sugar-phosphate backbone of each strand of a DNA molecule contains negative charges due to the phosphate groups. A type of protein called *histones* has more than 20 percent of its amino acids positively charged, so they bind to the negatively charged backbones of the two DNA strands. This forms the first level of packaging called a nucleosome: a beadlike structure made by a segment of the DNA molecule wound around a protein core that consists of two copies of each of four kinds of histones: H2A, H2B, H3, and H4. The N-terminus of amino acids called the histone tail sticks out of the nucleosome beads.

Chromatin fiber (30 nm structure). The DNA string between two consecutive nucleosomes is called a *DNA linker*. A histone called H1 binds to the linker, which begins the second level of packaging. Histone tails of one nucleosome interact with the DNA linker and the nucleosomes on both strands. This interaction causes the string of nucleosomes (the DNA molecule with proteins) to twist or fold resulting in a fiber of about 30 nm in diameter called chromatin fiber.

Looped domain (300 nm structure). Due to further interactions between different parts of the long chromatin fiber (with a 30 nm diameter), the fiber forms multiple loops, which are collectively called *looped domain*. These domains are attached to a scaffold which contains largely nonhistone proteins such as *topoisomerase*. The result of these interactions is a fiber that is about 300 nm in diameter.

Chromosome (700 nm structure). The looped domains interact with each other to further compact the DNA-protein structures (chromatins) to what is called a chromosome, a structure with a width of about 700 nm. Therefore, a replicated chromosome (with two DNA molecules) has a width of about 1400 nm.

But where does DNA comes from? We get it from our parents and then replicate it as needed.

9.4 DNA Replication

DNA replication is the process of making a copy of the existing DNA molecule. The DNA replication mechanism is built into DNA structure in terms of the fact that the two strands of the DNA molecule are complementary to each other. As illustrated in Figure 9.5, each strand of the existing molecule makes a new complementary strand and binds to it to make the double helix molecule Thus, this process replicates the original molecule into two molecules, each with one original strand and one new strand. This is called the semi-conservative model, which has been experimentally verified. A number of proteins work in the replication process illustrated in Figure 9.6 and are described in the following:

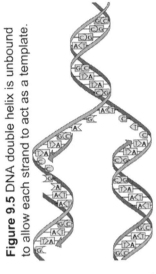

Figure 9.5 DNA double helix is unbound to allow each strand to act as a template.

1. Replication begins at sites called the origins of replication.

2. Proteins that recognize the nucleotide sequence at the origin bind to it and separate the two strands to open them up, forming a shape called a replication bubble. The protein *helicase* unwinds and separates the strands of the parent DNA molecule for replication to occur.

3. From the replication bubble, the replication begins in both directions, forming two replication *forks* as both strands act as templates to form new strands: one template for each new strand.

Molecular Genetics

> Complementary strands. The two strands of a DNA molecule are called complementary strands because each of them is the counterpart of the other. For example, if one runs in 5'-3' direction, the other runs in the 3'-5' direction; and along the length where one has A, the other has T, and where one has C, the other has G.

4. A protein called *topoisomerase* rejoins DNA strands after breaking and swivels them in order to relieve the strain caused by overwinding ahead of the replication fork.

5. A protein called a single stranded binding protein binds to the unwound single strand and stabilizes it until it can be used as a template for replication.

6. Two important factors that drive replication are initiation and elongation:

Initiation. The replicating protein (*DNA polymerase III*) cannot start building a complementary strand from a template from scratch; it can only add nucleotides to an already existing chain of nucleotides. So, an RNA polymerase called *primase* synthesizes a short RNA sequence called primer from the template starting at the 3' end, and then DNA polymerase III begins adding nucleotides to it that are complementary to the template.

Elongation. DNA polymerase elongates the new strand by adding nucleotides one by one only in the 5'-3' direction. This, added to the fact that the strands of the template run antiparallel, gives rise to the leading strand and lagging strand. The leading strand is synthesized continuously, whereas the lagging strand is synthesized in pieces called Okazaki fragments, which are sealed together by a protein called *DNA ligase*.

7. DNA polymerases proofread the DNA sequence and repair the mismatching (of bases forming base pairs) between the template and the newly built strand.

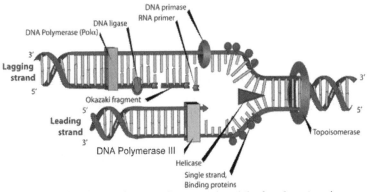

Figure 9.6 Synthesis of the leading strand and the lagging strand during DNA replication

8. *DNA polymerase I* removes RNA nucleotides which were used as a primer to begin replication of the template.

9. Proteins called nucleases identify and remove or excise incorrectly placed nucleotides, and the gap is filled with correct nucleotides, a process called *excision repair*. The errors that escape repair amount to only one in 10 billion nucleotides. These errors are called *mutations*: most of them are harmful for organisms, but once in a while a useful mutation occurs. Mutations are the root cause of evolution.

The replicated DNA molecule will be identical to the parent DNA molecule, but the replicated strand will be complementary (not identical) to the parental strand, as illustrated in a simple example in the next section.

Molecular Genetics

9.5 DNA Replication: An Example

In this section, let us walk through an example of DNA replication.

Build the replicated (new) DNA strand from the following parental strand, Strand 1, as the template:

5'- A-T-T-T-C-G-A-G-G-C-C-T-A-T-T-C-G-G-T-C-C-G-3'

Recall that the new strand will be complementary to the template strand. Remember:

- The complement of A is T and vice versa.
- The complement of C is G and vice versa.
- 5' is complementary to 3' and vice versa.

Applying these rules to the parent strand results in the following:

5'-A-T-T-T-C-G-A-G-G-C-C-T-A-T-T-C-G-G-T-C-C-G-3'
(parental strand, Strand 1)

⬇

3'-T-A-A-A-G-C-T-C-C-G-G-A-T-A-A-G-C-C-A-G-G-C-5'
(replicated strand, Strand 3)

Strand 3, which is the replicated strand, is identical to Strand 2, the other strand of the original DNA molecule. The replication of Strand 2 will result in another strand, Strand 4. Convince yourself that Strand 4 will be identical to Strand 1. Now, if you combine Strand 1 with Strand 3 and Strand 2 with Strand 4, you get two DNA molecules that are both identical to the original molecule (Strand 1 and Strand 2).

Because it is inherited, DNA and its products should have documented the hereditary background of organisms and species and hence provide evidence for evolution at a molecular level. A very important product of DNA is protein.

9.6 Central Dogma of Molecular Biology: From DNA to Proteins

DNA is the blueprint of life. Proteins, which are responsible for almost all dynamic functions of an organism, are made according to the instructions in DNA. Synthesizing a protein from the instructions in a gene of DNA is called *gene expression*. The mechanism of gene expression to make proteins from genes is called the *central dogma* of life, which has three main stages: transcription, migration, and translation. These stages are illustrated in Figures 9.7 and 9.8 and are described in the following.

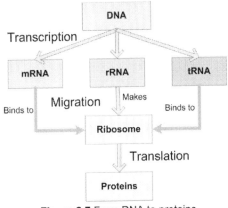

Figure 9.7 From DNA to proteins

Transcription. Just like a DNA polymerase makes a complementary copy of a DNA strand during replication, an RNA polymerase makes a complementary copy of a segment of DNA called a gene. This copying is called transcription in which nucleotide A is transcribed to U instead of T because RNA contains U instead of T. In other words, transcription is the gene-directed synthesis of RNA in the 5'-3' direction because just like DNA polymerase, RNA polymerase can also add nucleotides only to the 3' end of the RNA strand being developed. The DNA strand on which the gene under transcription exists is called the template strand or the noncoding strand, and the other DNA strand is called the coding strand. The transcript of a protein-coding gene is called messenger RNA (mRNA). Transcription is also used to express genes which do not code for proteins such as those genes which are transcribed into RNAs such as ribosomal RNA (rRNA) and translation RNA (tRNA). The rRNA combines with proteins to make ribosomes, the production sites for proteins.

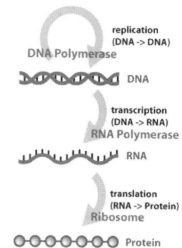

Figure 9.8 Central dogma of life

Migration or transportation. In a eukaryotic cell, transcription happens inside the nucleus. The rRNAs combine with proteins imported from the cytoplasm to make what are called small ribosomal subunits and large ribosomal subunits. The small and large subunits exit the nucleus through nuclear pores in the membrane, and so do mRNA and tRNA.

Caution! Transcription is the process that is used to synthesize several different types of RNA, not just mRNA, from a DNA template strand.

Translation. Translation is the process of synthesizing a polypeptide at a ribosome according to the instructions in the mRNA. As illustrated in Figure 9.9, the base sequence of an mRNA is translated into the sequence of amino acids forming a polypeptide, which would subsequently fold into a protein. The basic building block in the mRNA sequence is a codon that consists of three nucleotides. Each codon corresponds to a unique amino acid. As mRNA moves through the ribosome, a tRNA brings to the ribosome the amino acid corresponding to the codon at the ribosome. This way codons are read one by one and the corresponding amino acids are added to the growing polypeptide chain at the ribosome.

Let us take a look at the codon.

9.7 Codon: A Building Block

A codon is a triplet of bases representing a sequence of three nucleotides in mRNA and codes for a specific amino acid. The mRNA is a sequence of codons. Recall that nucleotides are the building blocks to make an information storage molecule, DNA, and the information transfer molecule, mRNA. Amino acids, which are coded in codons, are the building blocks to make proteins, which are the work horses in the cells.

So, codons are the units of information transfer from DNA to mRNA and subsequently to a protein. With only a few exceptions, a codon represents a specific amino acid. In that sense, codons are basic units of genetic code: if A, C, G, T (or U) are the letters of the genetic code, codons can be considered words. The mRNA that is synthesized from one strand of DNA is called the template strand. The mRNA strand is the complement of the DNA template strand. Let us illustrate with an example:

1. Consider a DNA template strand for mRNA:

    3'-AATTATCCGACG-5'

2. Rewrite the strand in the form of triplets:

    3`- AAT TAT CCG ACG -5`

3. Transcribe the strand; that is, write the complement of each triplet, letter by letter:

Molecular Genetics

5'-AAT TAT CCG ACG- 3' (DNA)

3'-UUA AUA GGC UGC- 5' (mRNA)

Note that T transcribes into A, whereas A transcribes into U and not T because RNAs have U instead of T. So, the codons in the mRNA corresponding to the given DNA strand are UUA, AUA, GGC, and UGC. Note that the complement of 5' is 3' and vice versa.

Problem 9.1

A. Given the genetic code as it is, how many codons can possibly be made?

Solution:

The total number of bases (nucleotides) from which to make codons = 4; they are A, T, C, and G.

Also, a codon is a triplet, and it has three bases.

Maximum number of codons = Maximum number of unique triplets that can be made from four different items =

$4^3 = 64$

B. If each codon codes for an amino acid, how come we only have 20 amino acids coded by 64 codons?

Solution:

Multiple codons (but very specific ones) can code for the same amino acid. So, in general, the relationship between codons and amino acids is many to one. Also, as explained in Section 9.8, 3 out of 64 codons do not code for any amino acid.

Molecular and Cellular Biology

As Problem 9.1 demonstrates, you can make a total of 64 unique triplets out of four types of nucleotides represented by A, U, G, and C. As shown in the next section, only 61 out of these 64 codons code for amino acids.

9.8 The Dictionary of Codons

	Second base in the codon				
First base in the codon: 5' end of codon	**U**	**C**	**A**	**G**	**Third base in the codon: 3' end of codon**
U	UUU ⎫ Phe UUC ⎭ UUA ⎫ Leu UUG ⎭	UCU ⎫ UCC ⎬ Ser UCA ⎪ UCG ⎭	UAU ⎫ Tyr UAC ⎭ UAA ⎫ Stop UAG ⎭	UGU ⎫ Cys UGC ⎭ UGA Stop UGG Trp	U C A G
C	CUU ⎫ CUC ⎬ Leu CUA ⎪ CUG ⎭	CCU ⎫ CCC ⎬ Pro CCA ⎪ CCG ⎭	CAU ⎫ His CAC ⎭ CAA ⎫ Gln CAG ⎭	CGU ⎫ CGC ⎬ Arg CGA ⎪ CGG ⎭	U C A G
A	AUU ⎫ AUC ⎬ Ile AUA ⎭ AUG Met or Start	ACU ⎫ ACC ⎬ Thr ACA ⎪ ACG ⎭	AAU ⎫ Asn AAC ⎭ AAA ⎫ Lys AAG ⎭	AGU ⎫ Ser AGC ⎭ AGA ⎫ Arg AGG ⎭	U C A G
G	GUU ⎫ GUC ⎬ Val GUA ⎪ GUG ⎭	GCU ⎫ GCC ⎬ Ala GCA ⎪ GCG ⎭	GAU ⎫ Asp GAC ⎭ GAA ⎫ Glu GAG ⎭	GGU ⎫ GGC ⎬ Gly GGA ⎪ GGG ⎭	U C A G

Figure 9.9 Standard genetic code: translation of the mRNA codons to amino acids.

The genetic code was cracked by a series of experiments performed in the first half of the 1960s. As illustrated in Figure 9.9, it was determined that only 61 out of 64 codons code for 20 amino acids and that more than one amino acid can code for the same amino acid.

The codons that do not code for an amino acid are UAA, UAG, and UGA, which instead code for *stop* signals that mark the end of translation from mRNA to protein. Another codon, AUG, which codes for the amino acid called methionine (Met), also functions as a *start* signal for the translation from mRNA to protein to begin. For this reason, the synthesis of each protein always begins with the amino acid methionine. However, some enzymes remove this first amino acid from their polypeptide chain.

As already described, the process of synthesizing codons (or mRNA) from DNA is called transcription. Let us explore transcription in further detail.

9.9 Transcription, Splicing, and Alternative Splicing

As illustrated in Figure 9.10 and explained below, there are three steps in transcription: initiation, elongation, and termination.

Initiation. In order to initiate transcription at a nearby gene, the RNA polymerase binds to a specific sequence of nucleotides in a DNA called a promoter. It is the promoter that determines which strand of the DNA helix is used as a template to express the gene. In bacteria, the RNA polymerase detects the promoter and binds to it, whereas in eukaryotes, the RNA polymerase II binds to the promoter only after a set of proteins called *transcription factors* bind to the promoter. The RNA polymerase II and the transcription factors bound to the promoter are collectively called a *transcription initiation complex*. Transcription factors are regulatory proteins that influence transcription, for example, by activating it or repressing it.

Molecular and Cellular Biology

Elongation. As illustrated in Figure 9.10, as RNA polymerase moves downstream along the template DNA strand in the 3'-5' direction by unwinding the DNA, it keeps adding nucleotides to the RNA strand (under construction) and hence elongating it in the 5'-3' direction. As the polymerase moves down the DNA strand, the two DNA strands behind it re-form the double helix. Multiple molecules of RNA polymerase can work on the same gene simultaneously, one following the other, to produce multiple transcriptions.

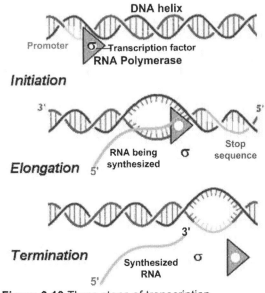

Figure 9.10 Three steps of transcription.

Termination. After transcribing the termination (or stop) sequence along the DNA strand, the RNA polymerase detaches from the strand, and the RNA transcript is released. This is called the primary transcript, and for bacteria this is the final transcript and is called mRNA. For eukaryotes, this is called pre-mRNA and is processed further to get mRNA, as illustrated in Figure 9.11.

The pre-mRNA contains noncoding segments called *introns* which are not to be used for gene expression. The segments to be used for gene expression are called *exons*. During processing, the 5' end gets a 5' cap, which is a modified guanine (G) nucleotide; introns are cut out, and exons are spliced together; and a *poly-A tail* is added to the 3'

Molecular Genetics

end. The process of cutting out introns and pasting exons together is called RNA splicing. This process is facilitated by a structure called *spliceosome*, which is an assembly of snRNAs (small nuclear RNAs) and proteins. A region of DNA from the promoter to the termination (stop) point that is transcribed into RNA is called a transcription unit. Depending on which segments in a primary transcript (pre-mRNA) are treated as exons and which as introns, RNA splicing can be performed in multiple possible ways called *alternative splicing,* giving rise to multiple mRNAs corresponding to the same pre-mRNA and hence the same transcription unit. For example, alternative splicing of the same transcription unit in a DNA can give rise to different but related proteins for the same type of tissues such as muscles. Because it generates variations, alternative splicing is considered to have played a significant role in evolution. For example, the sex differences in fruit flies (male and female) are largely due to the differences in the RNA splicing performed by males and females. Also, alternative RNA splicing partly explains how humans are doing quite well with relatively less number of genes: not even twice as many as a fruit fly.

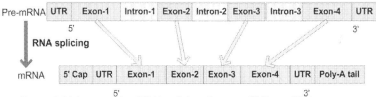
Figure 9.11 An example of RNA splicing of a pre-mRNA to mRNA

Once transcription produces mRNA, it can be translated into a polypeptide chain. Let us now explore some details of this translation.

9.10 Translation of mRNA Into a Polypeptide Chain

The process of translation is performed by three RNAs: mRNA which contains the message in terms of three-nucleotide long codons; tRNAs which translate the chain of codons into a chain of amino acids (one amino acid corresponding to each codon); and rRNAs which offer the site for this translation in the form of a ribosome. A ribosome has three sites on it which are used for translations: A, P, and E sites. Just like transcription, the process of translation also goes through three stages: initiation, elongation, and termination, which is illustrated in Figure 9.12 and described in the following.

> Anticodon. An anticodon is a nucleotide triplet at one end of a tRNA molecule. It base pairs with a specific nucleotide triplet on a mRNA called codon. This means that the anticodon on the tRNA is complementary to the codon on the mRNA with which it forms a base pair.

Initiation. Initiation itself goes through three steps. First, a small ribosomal subunit binds to an mRNA. Second, a tRNA, which has the anticodon UAC on one end and carries the amino acid corresponding to AUG (Met) codon on the other end, base pairs through hydrogen bonds with the AUG codon on the mRNA. Third, a large ribosomal subunit arrives and induces the tRNA with Met amino acid to attach to the P site. These three steps initiate translation. Now, the initiation site, which is called the A site, is available for the next tRNA to arrive. There are different kinds of tRNAs to bring different kinds of amino acids from the cystol to the ribosome.

Molecular Genetics

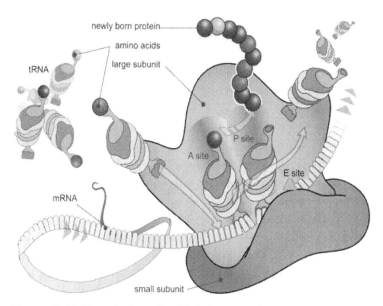

Figure 9.12 Translation of mRNA into a polypeptide.

Elongation. Like initiation, elongation consists of three steps. First, a tRNA of the appropriate kind with the right anticodon and amino acid comes along, and its anticodon base pairs with the next mRNA codon on the A site. Second, the amino acid attached to the tRNA on the A site makes a covalent bond (a peptide bond) with the amino acid on the P site. Now, the tRNA on the A site has the chain of amino acids. Third, the tRNA on the P site moves to the E site where it will be released, and the tRNA on the A site moves to the P site, making the A site available for the next tRNA. This third step is called *translocation*. This process is repeated for each codon on the mRNA and results in the addition of the corresponding amino acid to the polypeptide chain.

Termination. Elongation continues until a stop codon (UAG, UGA, or UAA) reaches the A site, at which point a protein called a *release factor* binds to the stop codon on the

Molecular and Cellular Biology

A site. A water molecule hydrolyzes (breaks) the bond between the completed polypeptide chain and the P site. This way, the last tRNA and the polypeptide chain are freed from the ribosome, and small and large ribosomal subunits depart.

> Caution! A given type of tRNA molecule can be attached to only one specific type of amino acid. So, there must be at least as many types of tRNA as the types of amino acids in its proteins. Moreover, because more than one codon can code for the same amino acid, the types of tRNA determined by the anticodon they bear should be larger than the types of amino acids available to make proteins.

An mRNA is usually translated by multiple ribosomes simultaneously as the mRNA passes through all of them producing multiple copies of a polypeptide chain. This cluster of ribosomes working on the mRNA is called a *polyribosome*.

9.11 Mutations

A mutation is a lasting change in the nucleotide sequence of an organism's DNA or in the RNA of a virus. Being the fundamental source of new genes, mutations are the primary reason for variations in traits of organisms and thereby play a significant role in evolution. But why do mutations occur? There is more than one reason for mutations to occur. For example, errors can occur during the replication of DNA, for instance, during cell division. A wrong nucleotide can be substituted for the right one, a nucleotide can be missed altogether, or an extra letter can be added to the sequence. These mutations are called *substitution, deletion,* and a*ddition* respectively. If these mutations involve only a single nucleotide, they are collectively called *point mutations*, which are changes in a single nucleotide pair of a DNA. Sometimes, however, a whole sequence (string) of letters can be reversed (an example of substitution), left out (deletion), or duplicated (addition).

> Caution! Only those mutations that occur in sex cells, which are also called germ cells or gametes, can be passed on from one generation to the next. In other words, mutations in the somatic cells of organisms that have a dedicated germ line such as those of animals are not transmitted to the descendants.

Another cause for mutations is the exposure to harmful environmental agents such as high energy radiation that can break chromosomes apart into pieces. Some pieces will be lost, and this will give rise to deletions of A, C, G, or T in the DNA string. Radiation can knock electrons right out of the atoms and generate a trail of free radicals, which can damage DNA. Other causes include mutagenic chemicals and viruses. Hermann Muller (1890-1967), an American geneticist, was awarded the 1946 Nobel Prize in Physiology or Medicine, "for the discovery that mutations can be induced by X-rays."

Most of the mutations are harmful to the organism in which they occur and to the offspring to which they are transferred. Figure 9.13 presents two examples of how mutations occur at a molecular (genetic) level and how their effects flow to the organism level through protein production.

Molecular and Cellular Biology

Figure 9.13.a shows part of the gene that codes for the production of the protein called hemoglobin, which

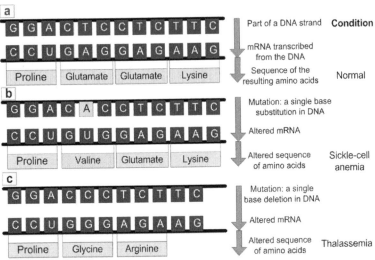

Figure 9.13 Examples of gene mutations and their effect on protein production and the organism

transports oxygen from lungs to other parts of the body. Figure 9.13.b shows the substitution of a single base A for T. As a result, a codon of the mRNA produced during transcription is altered, which in turn results in the production of valine amino acid instead of the glutamate amino acid for the polypeptide chain of the protein. This insertion of one wrong amino acid into the protein molecule causes a condition called sickle-cell anemia. Similarly, Figure 9.13.c illustrates how the deletion of a base, T, changes all the codons to the right and the corresponding amino acids. This results in the condition called *thalassemia*.

Damage or modifications caused to genes in the DNA by the environmental agents discussed earlier is passed on as mutations during the replication process, and these mutations last.

Molecular Genetics

Here are some facts discovered about mutations that are supported by data:

♦ The most common kind of mutation is the substitution in which a nucleotide of DNA is replaced with another.

♦ Each nucleotide in each gamete for each generation has one chance in a billion to be replaced by another nucleotide.

♦ Most mutations that occur in a random way are harmful (think of typos in your email message or in a computer code; you don't expect them to be good), whereas only a tiny fraction of them are beneficial to the fitness of the organism. Some of them, which are called neutral, are neither beneficial nor harmful.

♦ Only harmful and beneficial mutations play a role in evolution. Individuals with beneficial mutations are selected in and multiply over generations through reproduction, while individuals with harmful mutations are selected out and vanish along with their harmful (unfit) traits .

♦ Scientists have also observed that certain types of mutations occur more frequently than others. For instance, the changes in certain types of large sequences of bases occur more frequently than the average rate of overall single base-pair substitutions. As another example, *homopolymers*, continuous strings of eight or more identical letters, are especially prone to copying errors during DNA replication. It is the same case with sequences of two or more bases repeated over and over; these DNA regions are called *microsatellites*.

Molecular and Cellular Biology

> **Note.** Mutations in DNA and random recombinations during meiosis are major sources of genetic variations resulting in trait variations.

Mutations in genes that make proteins will affect protein synthesis. Do the cells make proteins continuously? You can guess that they do not because doing that would be similar to cell continuous cell division, which is a cause of cancerous conditions. Then how do they know when to make proteins and when not to make them? Well, they have a control system to regulate gene expression just like they have a control system for cell division.

9.12 Regulation of Gene Expression

Both prokaryotic and eukaryotic cells have an intrinsic control system to regulate gene expression. Furthermore, the system is dynamic enough to allow the cells to vary the pattern of gene expression in response to changes occurring in the environment.

In the following, we describe the gene expression control systems for bacteria (one of the two domains of prokaryotes) and for eukaryotes.

9.12.1 Regulation of Gene Expression in Bacteria: The Operon Model

In bacteria and phages, multiple genes are clustered together in a self-regulated structure called an *operon*, which includes an operator and a promoter in addition to genes. Because all the genes in an operon are under the control of a single promoter and operator, they are turned on and off collectively. This kind of control is called coordinate control. To illustrate the process, consider an example of a trp (tryptophan) operon, one of many operons in *E. Coli* living in the human colon. The bacterium needs the trp amino acid to survive, and it takes it from its environment. The bacterium produces it only when it is lacking in its environment such as when the human host is not eating. An operator is a DNA segment (a sequence of nucleotides) in and near the start of an operon to which a repressor binds in order to prevent the binding of an RNA polymerase to the promoter. The operator can be within the promoter or in between the promoter and the genes.

The process for the controlled production of trp amino acid in *E. Coli* is illustrated in Figure 9.14 and is described in

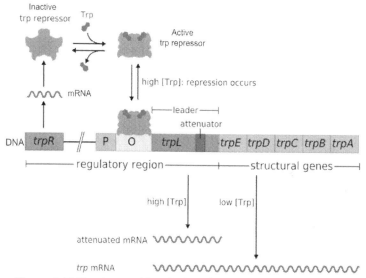

Figure 9.14 Structure and functioning of a trp operon with trp repressor

the following:

1. **Producing an inactive trp repressor.** The trp repressor protein is produced by expressing a regulatory gene called the trpR gene, which is located away from the trp operon and has its own promoter. The trp repressor is continuously produced, and it is produced in its inactive

form. In general, a repressor is a protein that when activated binds to an appropriate place on a DNA to inhibit gene transcription.

2. **Producing trp.** By itself (its default state), the trp operon is on because the trp repressor is in its inactive form. Therefore, the RNA polymerase can bind to the promoter and transcribe the genes in the operon to produce trp amino acid.

3. **Activating repressor.** When there is enough trp amino acid, some of its molecules bind to the trp repressor and activate it.

4. **Repressing trp production.** The activated trp repressor protein binds to the operon control (operator), which prevents the binding of the RNA polymerase to the promoter so that no mRNA is synthesized. This obviously inhibits the expression of the trp genes. Hence, the production of trp amino acid stops.

The control process that we have just described works by controlling when to prevent RNA polymerase from binding to the promoter and begin transcribing. As illustrated in Figure 9.14, a second level of control in the form of the trpL gene is coded for the so-called *leader peptide* and is followed by a sequence called an attenuator, which can provide a provisional stop signal if the RNA polymerase is already bound to the promoter and has begun transcription. The trpL and the attenuator are collectively referred to as the *leader region*, which is located just before the first gene to be transcribed. Attenuation is based on the fact that transcription and translation begin simultaneously in bacteria.

There are two types of operons: repressible operons and inducible operons.

Repressible operons. The trp operon is an example of a repressible operon, which is on by default and can be repressed. This type of operon is usually anabolic; that is, it builds an essential organic molecule. The repressor protein produced by a regulatory gene is inactive when produced. The molecule being produced from the expression of the genes of the operon bind to the repressor protein to activate it. The activated repressor protein then binds to the operator and turns off the operon.

Inducible operons. An inducible operon is an operon that is off by default and can be turned on. This type of operon is usually catabolic; that is, it breaks down food molecules for energy. In this case, the repressor protein produced by the regulatory gene is active and therefore binds to the operon operator and turns the operon off. The molecule that turns the operon on is called an inducer, which binds to the repressor protein and inactivates it. The *lac* operon is an example. Here is how it works. To start with, the regulatory gene called lacI, which exists outside the lac operon, produces the lac repressor protein, which is active. LacI binds to the operator of the lac operon and turns the lac operon off. When lactose (milk sugar) becomes available, one of its molecules called allolactose binds to the repressor protein and inactivates it, which turns the operon on. The operon then expresses its genes, which produce the enzymes used to metabolize lactose. This way, these enzymes are produced only when needed.

9.12.2 Regulation of Gene Expression in Eukaryotic Cells

In a multicellular organism, although different cell types have the same genome (set of genes), they express different subsets of genes. This is called differential gene expression, which leads to different cell types and organs necessary for the structure and functioning of a multicellular organism. Gene expression of a protein-making gene is measured by the amount of functional protein that is synthesized based on the gene code. Because there are several steps involved in reading the gene code and making the functional protein, the gene expression can be controlled at many stages from the packaging of DNA to the production and functioning of protein. In this section, we discuss the regulations at these different levels as illustrated in Figure 9.15.

> Genome. A genome is an entire set of genetic material in a single cell of an organism or in a virus. It includes coding and noncoding sequences. For example, the human genome consists of 23 pairs of chromosomes.

9.12.3 Regulation at the Chromatin Level

As explained in Section 9.3, DNA is packaged into chromatin by binding it to protein called histone. This packaging has various structural levels such as nucleosomes, chromatin fiber, and chromosome. As another example of the *form fits function* principle, this form or structure of chromatin supports regulation of gene expression. The more tightly and deeper the genes are bound into these levels of structures, the less likely they are to be expressed because they are less accessible for transcription. This relationship is largely expressed in the following two chemical interactions:

DNA methylation. This is the process in which a methyl group is added to the DNA molecule. Methylation causes DNA to be more tightly packed and therefore reduces gene expression. Methylation of genes is usually irreversible and is passed on to daughter cells during cell division.

Histone acetylation. This is the process in which acetyl groups attach to certain amino acids in the histone proteins. The acetylation causes the DNA in chromatin to be less tightly packed and therefore increases gene expression. To the contrary, histone deacetylation represses transcription.

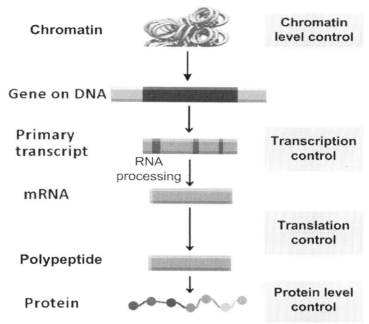

Figure 9.15 Different levels of gene expression control in eukaryotic cells.

9.12.4 Regulation at the Transcription Level

Each eukaryotic gene has its own promoter to which the RNA polymerase binds and begins transcription by moving downstream as explained in Section 9.9. As described in the following, there are three main control elements to regulate gene expression at the transcription level:

Transcription factors. This is a set of proteins that assist the RNA polymerase to initiate transcription. There are two kinds of transcription factors: general and specific. General transcription factors are essential for and common to the transcription of all protein-coding genes. Some of them bind directly to the promoter just like RNA polymerase does, others bind directly to the RNA polymerase, and some of them bind to each other. The transcription level is very low only if the general transcription factors are present in the cell. The specific transcription factors are the transcription factors which regulate (enhance or repress) transcription by interacting with specific control elements called enhancers in the DNA. In other words, each specific control element (nucleotides on the DNA) has its own specific transcription factor that binds to it. There are two kinds of control

Figure 9.16 Action of distal control elements (enhancers), activator proteins, mediator proteins, and promoter in a DNA.

Even though enhancers are far away from the promoter, the actions of the bending protein, specific transcription factors (activators), and mediator proteins enable enhancers to interact with and influence the promoter and hence the transcription.

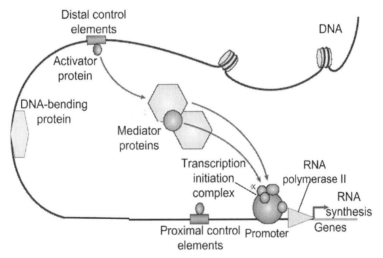

elements: proximal and distal.

- **Proximal control elements.** These are the DNA sequences located near the promoter. Some biologists consider them to be part of the promoter.

Molecular Genetics

- ♦ **Distal control elements.** Also collectively called enhancers, these are the DNA sequences located far from the promoter. As illustrated in Figure 9.16, when a specific type of protein called activators (specific transcription factors) bind to enhancers, a bending-protein bends the DNA strand so that it makes a kind of U-turn. This U-turn allows the activators bound to the enhancers to now be adjacent to the promoter. Subsequently, these sequences and activators bind to the promoter by mediator proteins and hence to the general transcription factors already on the promoter. This whole system, including general and specific transcription factors along with the RNA Polymerase II, form an active transcription initiation complex.

The transcription of a gene is done more efficiently when specific transcription factors acting as activators are present for all of the enhancer control elements of the gene. Specific transcription factors can also act as repressors.

Transcription level control also includes controls at the RNA processing level such as RNA splicing discussed earlier in this chapter. RNAs are synthesized in the nucleus of a cell and are transported to the cytosol for translation. During this RNA transportation, the nuclear membrane controls gene expression by regulating the transport of mRNA through it into the cytosol (or cytoplasm).

After mRNA passes through the nuclear envelope into the cytosol, gene expression can be controlled at the translation level.

9.12.5 Regulation at the Translation Level

Some examples of regulation at this level are mRNA degradation and the blocking of translation initiation by regulatory proteins. Gene expression can be blocked by degrading an mRNA and hence preventing translation. The mRNAs in eukaryotic cells live longer than those in bacterial cells. However, bacteria use the short life span of their mRNAs (a few minutes) to their advantage. In other words, this feature has evolved. Bacteria can leverage their mRNAs to quickly change the pattern of protein synthesis in response to changed environmental conditions. On the other hand, eukaryotic cells use the long life span of their mRNAs (a few hours to a few weeks) to translate them repeatedly and enhance gene expression efficiency.

As another means of control at the translation level, some regulatory proteins can bind to specific sequences within the untranslated part of the mRNA and thereby prevent it from binding to the ribosome. This results in blocking the initiation of translation or in terminating translation.

9.12.6 Regulation at the Protein Level

Some examples of regulation at this level are controlling protein processing to make functional protein from polypeptides produced by translation and degrading a functional protein. Translation produces polypeptide chains, which usually go through a process in order to acquire a functional form. Furthermore, many polypeptide chains go through chemical modification which is controlled by regulatory proteins before they can become functional proteins.

If the cell cycle is to function properly, the lifespan of each type of protein must be strictly regulated. This regulation is called selective degradation, which marks a specific protein for degradation by attaching molecules of another protein called *ubiquitin* to this protein. Subsequently, special protein complexes called proteasomes, which contain enzymatic components, recognize this protein and destroy it.

Molecular Genetics

> **Think About It!**
>
> Consider a type of somatic cell in an individual that goes through regular cell division. Now assume that the gene making a type of protein that regulates the cell cycle has evolved a mutation that makes the protein unaffected by proteasomes.
>
> A. What type of disorder or disease could this mutation lead to?
>
> B. Would this mutation be transferred to the offspring of this individual?
>
> Answers:
>
> A. cancer
>
> B. no

9.13 In a Nutshell

Here are some crucial points of molecular genetics:

- DNA is composed of four types of nucleotides identified by their bases: A, T, G, and C.

- In a DNA molecule, adenine (A) pairs only with thymine (T), forming two hydrogen bonds, and guanine (G) pairs only with cytosine (C), forming three hydrogen bonds.

- In replicating a DNA molecule, DNA polymerase cannot initiate the synthesis of a polynucleotide chain; it can only add to an already existing chain. Therefore, the enzyme named primase synthesizes a small RNA chain using the parental DNA strand as a template. This RNA chain is called an RNA primer. The DNA polymerase starts building the new DNA strand by adding nucleotides to the 3' end of the RNA primer and by using the parental DNA strand as a template.

Molecular and Cellular Biology

- The complement of A is T when synthesizing DNA, whereas the complement of A is U when synthesizing RNA.

- Unlike DNA polymerase, RNA polymerase can start a polynucleotide chain from scratch.

- Because DNA polymerase can add nucleotides only to the 3' end of the new nucleotide chain, the new DNA strand elongates only in the 5'-3' direction.

- In the circular chromosome of bacteria, only one replication bubble arises, whereas in the linear chromosome of eukaryotes, multiple replication bubbles appear simultaneously.

- Each gene in a DNA molecule has a specific and unique sequence of nucleotides; it is this sequence that distinguishes it from other genes.

- RNA is composed of four different types of nucleotides distinguished by their bases: A, U, G, and C.

- Each polypeptide chain of a protein is a specific sequence of amino acids.

- The promoter, a sequence of nucleotides in a DNA molecule, determines which strand of the DNA helix is used as a template to express the gene.

- Unlike translation in eukaryotes, translation in bacteria can begin while transcription is still in progress.

- Unlike in bacteria, pre-mRNA in eukaryotes is the primary transcript from which the final transcript is made though RNA splicing.

- During RNA splicing, introns are thrown away, and exons are joined together to make the final mRNA.

- The primary transcript is complementary to the template strand and not the so-called coding strand.

Molecular Genetics

- Not all RNAs transcribed from DNA are translated into proteins.
- Each type of tRNA molecule translates a specific mRNA codon into a specific amino acid.
- The synthesis of a polypeptide chain always begins in the cytosol by a free ribosome. The appearance of a signal peptide causes the ribosome to bind to the ER membrane and continue synthesis there.
- Mutations are the fundamental source of new genes.
- Each cell type in a multicellular organism such as a human being contains the same genome (set of chromosomes and therefore genes) but expresses a different subset of genes.
- Regulation of gene expression is responsible for a fertilized egg developing into a fully functioning organism composed of many different cell types.
- Generally speaking, repressible enzymes work in anabolic pathways, whereas inducible enzymes function in catabolic pathways.
- Each cell type in a multicellular organism maintains a specific gene expression program that ensures that certain genes are expressed and others are not.
- Histone acetylation promotes transcription, whereas DNA methylation and histone deacetylation repress transcription. Methylation of genes is usually irreversible and is passed on to daughter cells during cell division.
- The regulation of transcription initiation depends on the interaction of specific transcription factors with specific control elements in enhancers.
- General transcription factors are equally applicable to all the protein-coding genes, but the transcription level is

Molecular and Cellular Biology

very low only if the general transcription factors are present in the cell.

♦ Specific transcription factors can also act as repressors.

9.14 Review Questions

1. Which of the following is a true statement about the DNA molecule?

   A. Adenine and thymine consist of two rings each and bond to each other with two hydrogen bonds.

   B. Guanine and cytosine are both double-ring structures and bond to each other with three bonds.

   C. Adenine and guanine are two-ring structures, whereas thymine and cytosine are one-ring structures.

   D. Adenine bonds to thymine, and guanine bonds to cytosine by using only one hydrogen bond.

2. In a DNA molecule, deoxyribose sugar bonds to a nitrogenous base and a phosphate group by using which bond?

   A. covalent bond

   B. hydrogen bond

   C. ionic bond

   D. Vander Wall bond

3. Which of the following is not true about pyrimidines?

   A. They consist of only one ring.

   B. They are nitrogenous bases.

C. A pyrimidine in one strand makes a hydrogen bond with another pyrimidine in the other strand of a DNA molecule.

D. A pyrimidine covalently bonds with a sugar in a DNA molecule.

4. During replication, a nucleic acid is synthesized in which direction?

   A. 5'-3'

   B. bottom-up

   C. 1'-2'

   D. 5'-1'

5. In a DNA molecule, thymine always bonds with ___, and cytosine always bonds with ___.

   A. guanine...adenine

   B. uracil...guanine

   C. adenine...thymine

   D. adenine...guanine

6. An mRNA is synthesized from a DNA template strand by the process called ___, and the mRNA is subsequently used to synthesize protein by the process called ___.

   A. transcription...gene expression

   B. transcription...translation

   C. translation...transcription

Molecular and Cellular Biology

D. translation...gene expression

7. The nucleic acid that is translated to make a protein is called ____.
   A. DNA
   B. rRNA
   C. mRNA
   D. tRNA

8. Which amino acid sequence will be generated based on the following DNA template strand?

   3'-CAA-AGA-GCG-5'

   A. Gln-Arg-Ala
   B. Val-Ser-Arg
   C. Arg-Ser-Val
   D. no amino acid will be generated

9. Heart cells and liver cells in a human body are different because ____.
   A. they have a different genetic code
   B. they have a different genome
   C. they have different sets of chromosomes
   D. different genes are expressed in them

Molecular Genetics

10. Eukaryotic cells control their gene expression by ____.
    A. DNA methylation
    B. histone acetylation
    C. mRNA degradation
    D. processing a polypeptide chain to form a functional protein
    E. all of the above

11. The nuclear envelope in a eukaryotic cell controls gene expression by ____.
    A. assembling ribosomes
    B. regulating the passage of mRNA into the cytosol
    C. regulating the passage of mRNA into the nucleus
    D. creating protein complexes called proteasomes

12. Spliceosome is a structure that plays a role in controlling gene expression by ____.
    A. destroying proteins
    B. degrading mRNAs
    C. facilitating RNA processing
    D. creating proteasomes

Molecular and Cellular Biology

13. The enzymatic complexes in the cytosol of a cell that break down a protein marked for destruction are called ___.

   A. ubiquitins
   B. nucleosomes
   C. proteasomes
   D. lactase

9.15 Answer Key

1. C.
2. B.
3. C.
4. A.
5. D.
6. B.
7. C.
8. A.
9. D.
10. E.
11. B.
12. C.
13. C.

Notes:

Q1. See Figure 9.3.

Q3. A pyrimidine makes a hydrogen bond with a purine (C with G and T with A) inside a DNA molecule.

Q8. The mRNA strand from this DNA strand is 5'-GUU-UCU-CGC-3', which according to the codon dictionary will generate Val-Ser-Arg (see Figure 9.9).

Chapter 10

Biotechnology

10.1 Biotechnology: The Big Picture

It has taken about 3.5 billion years of evolution for us humans and for the amazing diversity of life that we see around us today to emerge. The basis for this evolution is the trait variations caused by genetic variations. Gene mutations in DNA and recombinations during meiosis are natural sources of these genetic and hence trait variations. Actions and processes performed by us humans also result in new genes and combinations of alleles (genotypes) that give rise to new traits (phenotypes). The field that makes it possible to directly engineer heritable changes in cells to

Figure 10.1 Brewing, an early application of biotechnology. Image: Brewer designed and engraved in the sixteenth century by J. Amman.

Molecular and Cellular Biology, by Paul Sanghera
Copyright © 2015 Infonential.

Molecular and Cellular Biology

produce novel products including proteins and traits is called biotechnology. According to the *United Nations Convention on Biological Diversity*, biotechnology includes "any technological application that uses biological systems, living organisms, or derivatives thereof, to make or modify products or processes for specific use." Equivalently, biotechnology can also be defined as the study and manipulation of living organisms or their components such as organs, tissues, cells, and molecules for different purposes such as performing research and developing products and applications.

At the core of any definition of biotechnology is the use of living organisms by humans. This practice has been happening for centuries, arguably making biotechnology a centuries' old field. Such practices as selective breeding of plants and animals and using microorganisms to produce beer, wine, and cheese (Figure 10.1) are some early examples of biotechnology. In olden times, our ability to use organisms was limited by our eyesight. Today, scientific developments at the micro and nano scale have given rise to new fields in biology such as molecular biology and genetic engineering. This has opened the door to a revolution in biotechnology which is occurring right in front of our eyes and is tremendously widening the scope of biotechnology.

Biotechnology is an interdisciplinary field because researchers in biotechnology use knowledge, tools, techniques, and processes from many fields, including biology, computer science, chemistry, and physics. Of course, like any other field, it develops its own knowledge, tools, techniques, and processes as well.

At the most fundamental level of life are the genes in DNA molecules, which give rise to protein molecules by expressing themselves. Understanding and manipulating genes involve making copies of genes. Biotechnologists have developed tools for gene cloning, gene expression, and gene functions that we explore in this chapter. Genes express themselves according to the genetic code. We will also explore techniques to decipher the genetic code or DNA sequencing. Furthermore, we will explore only a few of the enormous number of products and applications being created by biotechnology.

10.2 Molecular Technologies: Core of Biotechnology

The field of biotechnology embraces life at all levels from molecules to cells to organisms. Molecules constitute and run cells, which are the fundamental structural and functional building blocks of life. Therefore, our experience with life at the macroscopic or organism level has its roots in the microscopic and nanoscopic worlds of cells and molecules. This is why molecular tools and techniques (technologies) such as DNA technologies make the core of biotechnology and benefit life at all levels through research and practical applications.

Molecular technologies are enabling scientists to study compelling questions and issues of life at the fundamental level, that is, at the molecular level. These questions cover life from the level of genes to proteins, cells, and organisms. The following are some examples of research objectives that address these questions and issues:

- Understand the structure and functions of genes including deciphering genetic code.
- Understand where, when, and under which conditions a specific gene expresses itself.
- Study exactly what role a given gene plays inside an organism.
- Study the differences of a specific gene in different organisms of the same species such as humans. One application of such a study is to link specific mutations or alleles of genes to genetic (or hereditary) disorders.

Molecular and Cellular Biology

♦ Understand the functional differences between the same types of proteins in different organisms in terms of differences in their genes that code for those proteins.

To study these questions and issues, scientists have developed molecular technologies that have tremendously helped to advance the field of biotechnology. In the following sections, we explore some of these technologies.

However, before we can study genes, the blueprints of life, we need to make copies of them. Therefore, we begin with exploring technologies used to replicate genes: gene cloning and polymerase chain reaction (PCR).

10.3 Gene Cloning: Tools and Techniques

If you want to study and manipulate something, you often need many copies of a desired entity. In other words, you should be able to reproduce the entity under study. In sexual reproduction, organisms such as animals and plants do not reproduce exact copies (clones) of themselves. A living organism is defined as a clone of another organism if its genome is a genetic replica of the other organism's genome. Cloning is fundamental to biotechnology because reproducing genetic replicas is often essential to study and manipulate living entities. Also, you may be interested only in one or a few specific genes and not the whole genome or a whole DNA molecule.

> **Fascinating Fact!** A human gene is not larger than a millionth part of a human DNA. Thus in gene cloning, scientists are able to control and manipulate a millionth part of a molecule.

Just like nonliving matter can be best manipulated by working at the atomic level, living entities can be best manipulated by working at the gene level. Scientists and engineers frequently make use of natural processes. Genes can be cloned automatically by a natural process by introducing them to the DNA of an organism. When the organism replicates its DNA, the introduced (foreign) genes will also be replicated. The new DNA is called recombinant DNA because it is a recombination of the genes from the original DNA and the foreign genes. In other words, a recombinant DNA is a DNA molecule that contains segments of genetic material from different sources. The *in vitro* molecular technique used to isolate and manipulate fragments of DNA is called recombinant DNA technology.

> In-vitro. This is the technique of performing a given experiment in a controlled environment outside of a living organism such as in a test tube.

Recombinant DNA technology is very important in biotechnology because it is used for gene cloning and this way facilitates the study of relationships between gene sequences and *phenotypic* consequences of those sequences. Subsequently, this relationship can be used to control the phenotypes. Recall that a phenotype is a visible (or detectable) physical and physiological character (or a trait) of an organism such as eye color determined by its genotype (set of alleles).

In the following, we list and discuss the key tools or components of gene cloning and recombinant technology.

Host organism. This is any convenient organism that is used as a host for the foreign genes, the genes that need to be cloned. The genes are cloned (replicated) as the host organism replicates (reproduces) itself. A commonly used host organism is a rod-shaped bacterium called *E. Coli*, which divides itself every 20 minutes by binary fission. In addition to its frequent reproduction, another advantage of using *E. Coli* is that, like some other bacteria, it has one or more plasmids in it.

Plasmid. A plasmid is a small circular DNA molecule with a few genes in it. It exists inside some bacteria, such as *E. Coli*, separate from the chromosome. The genes in plasmids are not required for the survival and reproduction of bacteria, even though they can be useful under certain environmental conditions. Nevertheless, as Figure 10.2 illustrates, they are replicated independently of the chromosome as the bacterium reproduces asexually through binary fission.

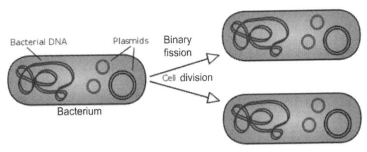

Figure 10.2 Replication of plasmids along with the replication of the cell or bacterium.

Plasmids serve as important tools in genetics and biotechnology labs where they are used for cloning specific genes such as vectors.

Cloning vectors. Also called a DNA vector, a cloning vector is a DNA molecule used to carry the foreign genes or DNA segment from its source to a target where it is replicated. There are commonly two kinds of vectors:

- **Viral vector.** This is a vector that uses a virus to carry the genes from a source to a target.

Biotechnology

- **Plasmid vector.** This is a bacterium plasmid carrying genes from a source to a target (host). The plasmid is isolated from the bacterial cell, and the foreign genes that need to be cloned are inserted into it. Plasmid vectors are used in genetic engineering, the application of scientific knowledge to manipulate genes for practical purposes, such as developing components, systems, and characteristics (traits). Figure 10.3 presents a commercially available plasmid vector.

Different parts of the vector in the figure are labeled to make you realize that scientists can identify and manipulate different parts or features of a vector. Discussion about these different parts is beyond the scope of this book.

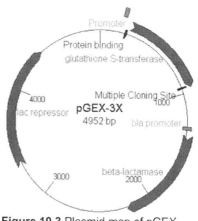

Figure 10.3 Plasmid map of pGEX-3X, a commercial cloning vector.

Restriction enzyme. This is an enzyme that recognizes and cuts a DNA molecule or strand at specific nucleotide sequences.

Restriction site. This is a specific nucleotide sequence of a DNA molecule or strand that can be recognized and cut by a restriction enzyme.

> **Think About It!**
>
> Q. Is a plasmid vector a recombinant DNA?
>
> A. Yes, it is a recombinant DNA because it contains genetic material from multiple sources.

Molecular and Cellular Biology

Bacterial plasmids are commonly used as cloning vectors for the following reasons:

- They can be easily isolated from bacteria.
- They can be easily manipulated to form recombinant plasmids by inserting foreign DNA into them.
- They can be easily reintroduced into the bacterial cells after they hold the foreign DNA.
- The recombinant bacterium plasmids replicate rapidly, which results in a high production rate.

The tools, techniques, and features discussed in this section are commonly used in methods of gene cloning.

10.4 Gene Cloning: The Process

There are many methods of gene cloning. However, all of them share some common features that we present through an example illustrated in Figure 10.4. As illustrated in this figure, here are the main steps for gene cloning using plasmids in the recombinant DNA technique:

1. Obtain the engineered vector DNA, which is also called plasmid DNA, originally fetched from an appropriate organism such as *E. Coli*.

2. Obtain the DNA of an organism whose genes need to be cloned.

3. Using the same restriction enzyme, cut the DNA fragment from both DNA samples. Now, the vector DNA has the site available for the DNA fragment to be cloned to fit in.

Biotechnology

4. Now both the cut plasmid DNA and the DNA fragment have sticky ends as shown in the figure. A sticky end is a single-stranded extension that can base pair with a complementary sticky end. Mix cut plasmids and DNA fragments. They will naturally join by base pairing. Add DNA ligase to the mixture, which will seal them together by catalyzing the base pairing. The structure obtained from the DNA fragment (with genes to be cloned) inserted into the cut part of the plasmid is called recombinant DNA.

5. Mix the plasmid vectors (recombinant DNA) with the host cells of an appropriate organism such as *E. Coli*. Some host bacteria (cells) will take plasmids in a process called *transformation*. If you used a virus instead of a plasmid as a vector to carry the DNA fragment, this process of inserting it into a host bacteria is called *transfection*.

6. The genes in the inserted DNA fragment are cloned and multiply as the host cells divide.

> **Think About It!**
>
> Q. The DNA fragment inserts itself into the cut part of the plasmid because the sticky ends of the DNA fragment and the cut plasmid bond with each other. What kind of bonds are those?
>
> A. They are hydrogen bonds.

The genes are cloned for uses in research, as tools, and in applications. Here are some examples of applications. By inserting cloned genes, bacteria can be altered for cleaning up toxic waste. Appropriate genes can be cloned to produce useful proteins such as human growth hormones and insulin. It is also used in gene therapy, recombinant vaccines, screening humans for diseases, and producing transgenic plants and animals.

Molecular and Cellular Biology

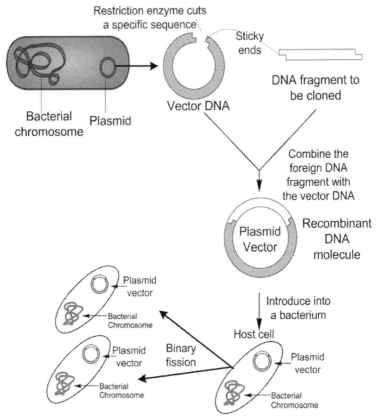

Figure 10.4 Gene cloning by using the recombinant DNA technique

One of the important uses of gene cloning is to prepare large quantities of a specific gene or a specific DNA segment for several uses. Another popular technique used to accomplish this is called polymerase chain reaction (PCR), which is discussed next.

10.5 Polymerase Chain Reaction (PCR)

Polymerase chain reaction (PCR) is a technique used to make copies of a DNA fragment in vitro with an exponentially amplified speed that is incredibly faster than the cloning method for making copies discussed in the previous section. PCR is a better option than cloning when the source of the genes that we need in large quantity is impure or scarce. The main components of PCR are described in the following:

Double-stranded DNA. This is one or more DNA molecules that contain the *target DNA segment* that needs to be amplified. These molecules act as templates.

DNA polymerase. This is an enzyme that accelerates the elongation of a new DNA by adding nucleotides to the 3' end of an existing DNA strand and thereby elongating it in the 5'-3' direction. DNA polymerase used in PCR must be heat resistant. A DNA polymerase commonly used is Taq DNA polymerase, which is thermostable. It is named after the thermophilic bacterium called *Thermus aquaticus* from which it was originally extracted in 1965.

Free nucleotides. These are the nucleotides, the monomers of nucleic acids, that a DNA polymerase adds to an existing DNA strand.

Primers. A primer is a short stretch of RNA with a free 3' end that is available for covalent bonding with a nucleotide. In PCR, short DNA strands about 20 nucleotides long can serve as primers.

In the beginning, template DNA, free nucleotides, and primers are mixed. During the PCR procedure on this mixture, which is illustrated in Figure 10.5, the cycles of reaction cause the target DNA fragment to exponentially grow in number of identical copies. The steps of a PCR run are explained in the following:

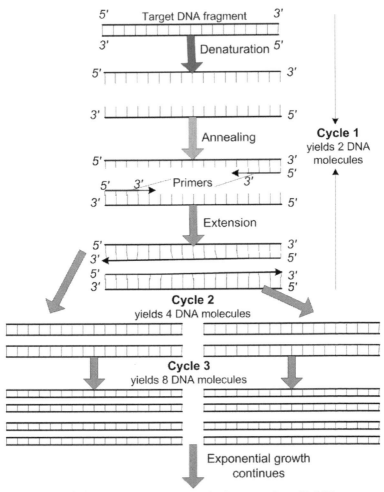

Figure 10.5 The polymerase chain reaction (PCR)

Initialization. Each PCR run is initialized once with a *hot start*, a process in which the temperature of the PCR mixture is raised to a high temperature, say about 95°C, and held there for about 10 minutes. In addition to facilitating denaturation for the first cycle, the hot start is aimed at other goals. For example, it increases the yield of the final product (desired DNA) by eliminating the undesired background. Also in the absence of the hot start, nonspecific binding of primers can happen. Furthermore, the hot start improves the PCR signal (ratio of desired product to undesired product) by eliminating side reactions which would happen at a low temperature. An example of a side reaction is the formation of *primer dimers*, that is, primers binding (base pairing) with (complementary) primers. This 10 minute long hot start is required to activate certain enzymes (polymerases) being used in PCR, and the exact temperature required for the hot start depends on the polymerase being used.

After the hot start, PCR goes through a number of repeated cycles. Each cycle progresses with temperature changes and goes through three steps: denaturation, annealing, and extension.

Denaturation. In PCR, this is the process of separation of two strands of a double helix DNA molecule. This is achieved by raising the temperature of the mixture to about 95°C and holding it there for about 20 to 30 seconds.

Think About It!

Q. The initial mixture prepared to start PCR contains greater amount of primers than that of DNA. Why do you think this is necessary?

A. The greater amount of primers is necessary to increase the probability of the primers binding to the DNA strands and decrease the probability of the originally separated DNA strands binding to one another.

Annealing. In general, this is the process of a sequence of nucleotides (DNA or RNA strand) base pairing with its complementary sequence by using hydrogen bonds to form a double-stranded polynucleotide. In PCR, it means primers forming hydrogen bonds with ends of target sequences, that

Molecular and Cellular Biology

is, separated DNA strands. This is accomplished by lowering the temperature of the PCR mixture to about 40-65°C following the denaturation step.

Extension. DNA polymerase begins adding nucleotides to the 3' end of each primer bonded to the target sequence by using the target sequence as a template. The temperature during this stage is usually raised, say to 65-72°C. In this process of extension or elongation, the DNA polymerase synthesizes a DNA strand complementary to the strand to which the primer was bonded. These two complementary strands then base pair to form a double-stranded DNA molecule. If we began with one molecule of DNA, we now have two molecules of the same DNA.

The mixture is heated again to denature, and the second cycle begins, which will yield four molecules from each of the two molecules. Similarly, the third cycle will yield 8 molecules from each of the four molecules, and this way the product (the number of DNA molecules) will keep growing exponentially as the cycles are repeated.

Problem 10.1

Assume a PCR run began with one molecule of interest. What is the maximum number of DNA molecules of interest it can produce if the PCR run is 30 cycles long?

Solution:

Number of DNA molecules produced = 2^{30} = 1073741824 $\cong$ 1 *billion*

PCR has enormous uses and applications, including amplifying a DNA sample obtained from a crime scene, quick diagnoses of genetic disorders, and detection of scarce and hard-to-detect viral genes. However, due to the errors in PCR, it is not a substitute for gene cloning, especially when a large quantity of exact copies is required.

The capability of obtaining large quantities of desired genes and DNA opened up doors to many applications in the field of research, enabling scientists to handle the challenging questions of molecular biology discussed in Section 10.2. Many of the technologies used to address these questions and issues use a common technique called gel electrophoresis, which we explore next.

10.6 Gel Electrophoresis

DNA molecules are the blueprint of life, and protein molecules are their product. Many techniques to study DNA and protein molecules include the use of gel electrophoresis, which is a technique to separate nucleic acids (DNA and RNA) and proteins based on their physical properties such as size and electrical charge. This technique makes use of physics laws according to which the motion of a particle depends on some of its physical properties such as size and mass in general and may also depend on its environment such as electric field.

The name *gel electrophoresis* comes from the fact that the molecules in this technique move through a semisolid gel under the influence of an external electric field. The gel commonly used in this technique is called *agarose*, a polysaccharide. As an example, assume we want to separate the DNA fragments of different lengths produced by a restriction enzyme or by PCR. We make this mixture of fragments move though the gel to which a known electric field has been applied. Under the influence of the electric field, the fragments move though the gel at different speeds depending on their sizes. The shorter a fragment is, the faster it will move. Fragments of the same size gather into one band.

Molecular and Cellular Biology

After the fastest molecule has made it close to the end of the gel chamber, the electric field is turned off. The molecules in the gel are stained so that they become visible under the appropriate wavelength of light such as ultraviolet light, and an image can be taken.

Figure 10.6 presents an image from a gel electrophoresis. Wells are the holes. In each of these holes, a sample (mixture) of DNA fragments of different lengths was placed. Each band in the figure represents DNA fragments of a specific length. Each band holds thousands of molecules of the same length.

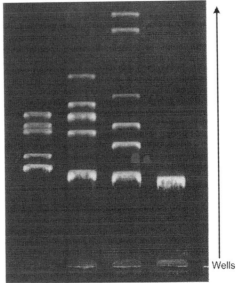

Figure 10.6 Image of gel electrophoresis

Information collected from the image is analyzed to learn about the DNA fragments under study. Gel electrophoresis is used in many other molecular techniques, including DNA sequencing.

10.7 DNA Sequencing

The genetic code in DNA is universal and is defined by the sequence of four types of nucleotides: A, C, G, and T; that is, it is defined by the order in which these nucleotides appear in a given strand of DNA. Therefore, in order to decode or decipher the genetic code of a DNA molecule, we need to determine its sequence. The process or technique that accomplishes this is called DNA sequencing. In this technique, a set of DNA strands complementary to a target (original) DNA strand are synthesized. These strands are terminated at different lengths.

The following are the main steps of the DNA sequencing technique:

1. A strand of DNA to be sequenced is mixed with the required ingredients for DNA synthesis: nucleotides of all four types (A,C, T, G), primer, and DNA polymerase. Also, four types of nucleotides tagged with different pigments (for example with fluorescent molecules) are added to the mixture. Each tagged nucleotide is designed to halt the synthesis of a DNA strand once added to it.

2. The DNA polymerase synthesizes a new DNA strand complementary to the original strand by randomly adding untagged (normal) or tagged nucleotides to it. Once a tagged nucleotide is added, the synthesis is terminated, and the first tagged nucleotide becomes the last nucleotide of the synthesized strand. Because polymerase is picking up a tagged nucleotide at random, the process will yield synthesized strands of different lengths. At the end, the strands with all possible lengths ranging from one nucleotide to the length of the original strand exist in the mixture.

3. The mixture containing the synthesized strands is run through gel electrophoresis, which separates the strands of different lengths into different bands. Each band contains strands of the same length tagged with the same color. This means that each band has a certain color uniquely belonging to one of the four nucleotide types. Two bands differing in the length of their strands by just one nucleotide can be distinguished from each other.

Molecular and Cellular Biology

4. A computer detects the colors and intensity (number of strands) of each band and creates an image (spectrogram) like the one shown in Figure 10.7. The color of each band represents the color and hence the identity of the last nucleotide in each strand of that band.

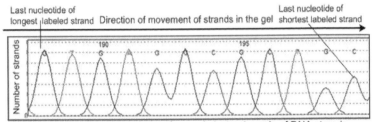

Figure 10.7 The fluorescent image from gel electrophoresis of DNA strands produced during the DNA sequencing technique.

5. A sequence (from 5' end to 3' end) is constructed by reading the spectrogram from the shortest strand to the longest strand as illustrated in Figure 10.7.

6. The sequence complementary to the sequence obtained in Step 5 is the sequence for the original strand.

As discussed in Chapter 9, genes express themselves according to their code. Biotechnology has tools and techniques to study different aspects of gene expression.

10.8 Tools and Techniques to Study Gene Expression

If a gene is expressing itself into the corresponding protein, it must be transcribing to mRNA. If we want to measure the expression of a specific gene in a specific part of an organism, we can look for the mRNA corresponding to this gene. There are two techniques available to make this detection: Northern blotting and reverse transcriptase-polymerase chain reaction (RT-PCR).

Biotechnology

Northern Blotting. Northern blotting is a molecular technique used to detect specific nucleotide sequences in a mixture of RNA. It can be used to measure the expression of a specific gene in a specific cell or tissue type by analyzing a sample of RNA from that cell or tissue type.

> **Note.** Northern blotting was named for its similarity to another technique called Southern blotting, which is used to detect specific DNA sequences among a mixture of DNA molecules instead of RNA molecules.

As illustrated in Figure 10.8, the following are the main steps of this technique:

1. **Extraction of RNA.** Total mRNA is extracted from a source, which could be a tissue or a cell.

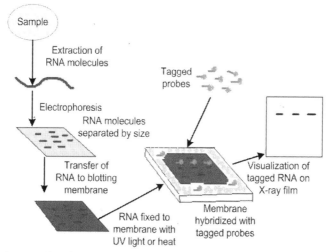

Figure 10.8 General procedure for detecting RNA by Northern blotting.

2. **Gel electrophoresis.** These samples of mRNA are separated by size by running them through gel electrophoresis.

257

3. **Transfer.** The separated mRNA samples are transferred from the gel to a blotting membrane. This step is the source of the name Northern blotting, even though the whole technique is called Northern blotting.
4. **Tagged hybridization.** A small piece of a DNA or RNA strand called a probe is tagged (labeled), for example, with a fluorescent dye or a radioactive atom. This probe is a sequence of nucleotides complementary to the sequence that we are searching for in the sample. This tagged probe is transferred to the membrane where it binds to its complementary strand in the sample if it exists. This process of binding is called hybridization.
5. **Visualizing the signal.** The tagged hybridized signal can be visualized or detected with X-ray film.

Another technique called reverse transcriptase-polymerase chain reaction (RT-PCR) can be used to compare gene expressions in different samples.

RT-PCR. Reverse transcriptase-polymerase chain reaction (RT-PCR) is a molecular technique used to compare gene expressions in different samples, for example, in different cell or tissue types or in the same cell or tissue type but under different conditions. As the name suggests, this technique uses an enzyme called reverse transcriptase and a technique called PCR, which is discussed earlier in this chapter.

> **Caution!** Do not confuse reverse transcriptase-polymerase chain reaction (RT-PCR) with real time polymerase chain reaction, another form of PCR technique, which is also sometimes referred to as RT-PCR.

The following are the main steps of RT-PCR:

1. **Extraction of mRNA samples.** Extract mRNA from each of the sources whose gene expression is to be compared.
2. **cDNA synthesis.** Synthesize complementary DNA (cDNA) by incubating each mRNA sample with reverse transcriptase and other required ingredients. You basically create the DNAs (genes) from which mRNAs were transcribed.
3. **PCR amplification.** Amplify cDNA by using PCR in which you use primers specific to the gene that you are searching for in the samples.
4. **Gel electrophoresis.** Run the product of PCR through gel electrophoresis. If the target gene does not express itself, gel electrophoresis will produce no band. If the gene was expressed in some samples, the properties of the bands such as distance covered will reveal the relative sizes of the amplified fragments and hence the relative rate of expression.

RT-PCR has various uses, which include diagnosing genetic diseases and studying the genomes of some viruses.

10.9 Tools and Techniques to Determine Gene Functions

As you learned in Chapter 3, almost all the dynamic functions of an organism stem from or depend on proteins which in turn are originated from genes. All the inherited traits of an organism have their roots in genes. Therefore, to understand organisms, it is necessary to determine gene functions.

Because genes express themselves through the proteins that originate from them, studying the protein coded in a gene is one way of studying a given gene or to determine its function. The study of proteins, including their structures and functions, is called *proteomics*. Another way to determine the function of a gene is to observe what happens to an organism when that gene is missing. There are various ways to perform this study. For example, study two similar organisms when one of them is lacking that specific gene, or study the same organism with the gene and then without the gene, and compare the results. In studying the functions of human genes this way, scientists often use mice because of all the model organisms used in the lab, mice have a genome that most closely resembles the human genome.

For all practical purposes, an organism can be considered lacking a gene if that gene does not express itself, even if the organism does have that gene. There are methods to silence an gene, that is, to stop it from expressing itself. One such technique is call RNA interference (RNAi). This technique uses a synthetic double-stranded RNA molecule that matches the sequence of the target gene. This RNA molecule interferes by either blocking the transcription of the gene or by breaking down the mRNA transcribed from the gene.

Yet another technique to determine the function of a gene is the comparative study of the genomes of a large number of organisms of the same species. For example, to determine which gene is responsible for a condition or a disease, you study the genomes of a large number of people with the same condition and see what gene or allele they share with each other which they do not share with the rest of the population. Such studies do not require complete DNA sequencing. Instead, scientists focus their study on the DNA sequences that vary among the population and are called genetic markers. The location of a genetic marker on a chromosome is known and therefore is easily identifiable.

Studying the genomes of organisms is emerging as a field in itself that is referred to as genomics.

10.10 The Rise of Genomics

A genome is the set of all chromosomes inside a cell of an organism. For example the human genome consists of 23 pairs of chromosomes, that is, 46 chromosomes total. Each chromosome has one DNA molecule in it which contains genes. Genomes vary widely in size across species: the smallest known genome for an autonomously living organism (a bacterium) contains about 600,000 DNA base pairs, whereas human and mouse genomes have about 3 billion base pairs. Each cell consists of a complete genome, that is, a complete set of genes for the individual, which means the entirety of the hereditary information for the individual. To understand an organism, it would be very important to sequence, that is, to decode its whole genome.

The sequencing of the human genome was completed in 2003 by both public and private efforts. The public effort was led by the U.S. National Institute of Health (NIH) in an undertaking known as the Human Genome Project, which began in 1990. So, in 2003, fifty years after the discovery of the double helix structure of DNA, the sequence of the 3 billion base pair long human genome was completed. Following this event, the genomes of many other species have also been determined through sequencing. Sequencing a genome determines where the genes are in the genome, not what they encode.

Sequencing has already produced an overwhelming amount of data. Sequencing is equivalent to taking things apart, and now scientists have begun to put things back together. This means they are trying to understand what the genes and non-gene parts in the DNA actually encode for and how different genes in a genome and different genomes within a species interact among themselves and how the different genomes of a species compare with one another. This study is called *genomics*. Informational technology (IT), including computation and internet technologies, is being used to produce, store, and analyze the biological data in genomics, and this has already given rise a to a field called *bioinformatics*, the application of information technologies to produce, store, retrieve, and analyze biological data. We already know the big molecular picture of life:

1. Gene sequence determines the protein sequence, that is, the sequence of amino acids in the polypeptide chain.
2. The protein sequence determines the protein structure, including its three-dimensional structure.
3. The protein structure determines the function of a protein.
4. Molecular regulatory mechanisms, mostly run by protein-gene or protein-protein interactions, govern and control the pattern of gene expression as well as the delivery of the right function in the right amount at the right place at the right time.

Genomics and bioinformatics are exploring the details of these aspects of life dynamics. The field of *structural genomics* focuses on the determination (or prediction) of the three-dimensional structure of proteins based on the corresponding gene sequences. The field of *functional genomics* focuses on determining the functions of genes and proteins coded by genes and the interactions among them.

10.11 Real-World Applications of Molecular Biotechnology

Real world or practical world refers to the world outside the research labs and academia. To begin with, the whole field of biotechnology itself is an application of science which includes the application of principles of different branches of science, including biology, chemistry, physics, and computer science. Molecular biotechnology, a relatively new subfield of biotechnology, is giving rise to a wide spectrum of practical applications in almost all areas of life including but not limited to agriculture, energy, environment, forensics, and medicine and health. Here are some examples:

Agricultural products. The selective breeding method of biotechnology has been used for thousands of years to improve agriculturally important organisms such as animals and plants. The development of new disease-resistant wheat varieties by crossbreeding the existing varieties is an example. Animals have also been crossbred to produce animal types with desired traits. Molecular biotechnology has enabled scientists to manipulate organisms at the fundamental level, the level of genes that encode most of the characteristics of organisms. By introducing appropriate genes into an existing organism, you can change its characteristics or introduce new characteristics without waiting for the next generation. This offers a quicker and more precise alternative to selective breeding and opens the door to even more possibilities than selective breeding offers. An organism whose genome contains one or more genes from the genome of another organism of the same or different species is called a transgenic organism. Such an organism is also called a genetically modified organism (GMO). A great number of crop plants have already been genetically modified, resulting in crops with desired traits such as resistance to disease and spoilage and ripening at the desired time. GMO foods are commercially available and are being used. Here are some of the benefits of genetic modification (genetic engineering) in agriculture: enhanced crop protection, increased crop productivity, improved nutritional value and flavor, and some environmental benefits. The risks of GMO food include antibiotic resistance, allergens and toxins, and potential gene escapes.

Gene therapy. The manipulation of genes in the cells of an organism to treat a disease is called gene therapy. Manipulation may involve replacing or introducing a gene. This way a genetic condition giving rise to the disease is corrected. For permanent effect, gene changes must be made

Molecular and Cellular Biology

at a fundamental level such as the adult stem cell level for a specific organ or tissue so that the changes become relatively permanent, and the changed gene or allele spontaneously multiplies through cell division. For example, if you make the genetic change in the stem cells of bone marrow, the change will become effective permanently for all the blood cells and the immune system cells of the patient.

Genetic profiles. This is the unique set of characteristics or genetic markers in the genome of an organism. It can be determined by using techniques such as PCR. It is not full genome sequencing. A genetic profile is also called DNA fingerprinting because it uniquely identifies an individual. It can be used in forensics and in determining paternity and the probability of developing an inherited disease.

Think About It!

Q. Is gene therapy a form of recombinant DNA technology?

A. Yes.

Pharmaceutical products. Like many ancient biotechnology techniques or fields, the pharmaceutical industry has been here before the rise of modern biotechnology (largely based on molecular and cellular biology) as well. Most traditional pharmaceutical drugs consist of small molecules that bind to their molecular targets in the patient's body to activate or deactivate specific biological processes. These molecules are usually manufactured through traditional synthetic methods of organic chemistry.

The contribution of biotechnology to the pharmaceutical industry is the rise of biopharmaceutical drugs, which consist of large biomolecules such as proteins which can address problems not easily resolved by traditional drugs. An example of an application of DNA recombinant technology is the manufacturing of human insulin, a protein, by injecting the corresponding gene into the bacteria *E. Coli.* Insulin is used to treat diabetes. As another example, human growth hormones, which are proteins manufactured by using techniques of biotechnology, are being used to treat dwarfism in children. Another protein called tissue plasminogen activator (TPA) is also being manufactured which dissolves blood clots and help prevent heart attacks. The uses of biotechnology in the pharmaceutical industry are so enormous that for some the two words *biotechnology* and *pharmaceuticals* are almost synonymous.

Diagnosis. Biotechnology has developed techniques to diagnose genetic diseases. One of the common tools in genetic diagnosis is the DNA probe, which is a single-stranded fragment of a DNA molecule searching for a complementary strand. If it runs into its complementary strand, it combines with it and sends back a signal such as a radioactive signal. If necessary, PCR is used to increase the amount of the target DNA in order to increase the probability of detection of the complementary strand if it exists. Many genetic diseases are being diagnosed by using this technique.

DNA techniques are also used to diagnose nongenetic diseases. For example, tagged (or labeled) DNA fragments (probes) are used to detect pathogens. A prime example of this is that scientists are now able to detect HIV by using an appropriate nucleic acid probe because the RNA genome of HIV has already been sequenced. This can be accomplished by taking the blood or tissue sample of the patient, amplifying the HIV signal by using RT-PCR, and detecting the signal by using the nucleic acid probe.

Regenerative medicine. The field of stem cell research, which uses biotechnology techniques, is giving rise to a whole new application area called regenerative medicine. Regenerative medicine is the emerging field of medicine that focuses on restoring the normal function of a diseased body by replacing or regenerating its diseased parts, such as cells, tissues, or organs. What do stem cells have to do with it? By definition, stem cells have the capability to differentiate themselves into specific types of cells such as heart cells and liver cells and therefore can be used to regenerate desired differentiated cells, tissues, and organs. A differentiated cell is a cell that has acquired its specific functionality. Neurons, skin cells, and red blood cells are examples of differentiated cells.

Regeneration is a fundamental property of life working not only from generation to generation but also at the individual level. For instance, your body heals itself from a cut or wound, which disappears in a week or two. We regenerate our skin every week and our liver about each year. Each type of tissue has its own turnover time depending on various factors including the workload endured by its cells. Nevertheless, the entire human skeleton is replaced in about 7 to 10 years. And all this happens spontaneously through cell division; no surgeries are involved.

Here are the three interrelated big questions: how does our body develop out of a single cell, the zygote? How does our body keep renewing or regenerating its parts all through its life? What are the fundamental processes and mechanisms in action at the molecular, atomic, and particle level that make this generation and regeneration possible? Finding the answers to these questions is equivalent to redefining medicine. Answering these questions will enable us to direct the body to regenerate itself by replacing diseased parts with healthy ones. This will potentially lead to cures for fatal disorders and diseases, such as Alzheimer's, cancer, diabetes, heart disease, Parkinson's disease, and spinal cord injuries.

10.12 In a Nutshell

- Biotechnology is the use or manipulation of any organism to create applications or products.
- Biotechnology on a macro scale has been used for centuries such as in selective breeding to create organisms with desired properties and in brewing.

Biotechnology

- The developments in molecular biology have given rise to molecular biotechnology or modern biotechnology, which is creating an enormous number of products and applications in all areas of life.

- As illustrated in Figure 10.9, biotechnology has developed all the way from DNA to protein along the flow of the central dogma of life. It has techniques that work at the DNA level, RNA level, protein level, or a combination of these levels covering transcription and translation.

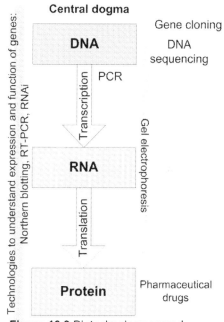

Figure 10.9 Biotechnology spread out along the central dogma

- Using biotechnology, we can clone specific genes and introduce genes from one organism into another to create products, improve the organisms, or to treat disorders.

- We can sequence genomes, that is, decode the genetic code, and we can study, monitor, and control the expression and function of a gene.

- In addition to creating products and applications, biotechnology is giving rise to whole new fields or subfields such as genomics, bioinformatics, and regenerative medicine, which have their own applications.

Molecular and Cellular Biology

10. 13 Review Questions

1. In gene cloning, a cut plasmid and a DNA fragment are joined at their ends by ____.

   A. DNA polymerase

   B. a restriction enzyme

   C. ligase

   D. primer

2. Which of the following is not a function of a hot start in PCR?

   A. improve the PCR signal

   B. formation of primer dimers

   C. eliminate the side reactions

   D. activate the polymerase

3. Why is the amount of primers in a PCR mixture higher than the amount of DNA?

   A. to ensure that the primers bind to the DNA strands rather than originally separated DNA strands binding to one another

   B. to ensure that DNA strands bind to complementary DNA strands and primers bind to complementary primers

   C. because a DNA strand needs more than one primer to bind to

   D. because primers are less stable than the DNA strands

4. Most of the activities of a cell are directly carried out by ____.

   A. genes that encode the proteins

B. proteins that are encoded in genes

C. carbohydrates

D. vitamins

5. Sequencing of a genome determines all of the following except ____.

   A. the chromosome that contains a specific gene

   B. the location of the gene on a DNA molecule

   C. the chemical code of a DNA molecule

   D. the specific function of specific genes, that is, what a gene does or encodes for

6. *E. Coli* is used to produce human insulin after the gene for the human insulin is introduced into it. Such an organism is called a ____.

   A. transgenic organism

   B. genetically modified organism

   C. selectively bred organism

   D. cloned organism

   E. A. and B.

7. Reverse transcriptase is used to ____.

   A. create an mRNA from a protein

   B. create cDNA from an mRNA

   C. compare the expressions of a gene under different conditions

   D. block the expression of a gene

   E. B. and C.

8. Which of the following is incorrect about a DNA probe?

Molecular and Cellular Biology

  A. It is a double-stranded DNA fragment.
  B. It is a single-stranded fragment of a DNA molecule.
  C. It is a tagged or labeled molecule designed to send a signal when a specific event occurs.
  D. It is used to detect specific DNA fragments in a sample.

9. Northern blotting is used to ____.
  A. clone genes
  B. study the function of a gene
  C. study gene expression by detecting the specific nucleotide sequences in a sample of RNA
  D. introduce genes from one organism into another

10. A technique used to separate molecules such as nucleic acids and proteins based on their physical properties is called ____.
  A. PCR
  B. gel electrophoresis
  C. RT-PCR
  D. Northern blotting

10.14 Answer Key

1. C.
2. B.
3. A.
4. B.
5. D.
6. E.
7. E.
8. A.
9. C.
10. B.

Notes:

Q2. A hot start helps prevent the formation of primer dimers, that is, primers bonding with primers.

Q5. Sequencing determines where the genes are in a genome and what their nucleotide sequences are, not what they encode.

Molecular and Cellular Biology

Glossary

activation energy. The minimum amount of energy that the reactants must absorb for the reaction to occur.

active transport. The transportation of an entity across the membrane against the electrochemical gradient of the membrane; energy expended.

Alternative RNA splicing. The process in which multiple mRNAs are produced from the same pre-mRNA, therefore enabling the production of different but related proteins from the same gene.

anabolism (or anabolic pathway). A metabolic pathway (set of chemical reactions) that uses energy to synthesize a complex molecule from simpler ones.

anticodon. A nucleotide triplet in a tRNA that base pairs with a complementary nucleotide triplet in an mRNA called a codon.

asexual reproduction. Reproduction in which the offspring inherits genes from only one parent.

acid. A substance that when added to a solution increases the hydrogen ion concentration of the solution.

atom. The smallest particle that represents a specific element.

ATP. Adenosine triphosphate molecules that store energy in a form that can be used by the cells to perform some functions.

autotrophs. Organisms that make their own food such as plants.

base. A substance that when added to a solution decreases the hydrogen ion concentration of the solution.

biochemistry. The chemistry of living entities.

bioinformatics. The field of biology focused on the application of information technologies to generate, store, retrieve, and analyze biological data.

biosphere. The sum total of all the ecosystems on our planet. This covers all parts of the Earth, including the environment, where organisms live.

biotechnology. Use of organisms or their components to produce practical applications or products.

buffer. A substance that resists the change to pH in a solution.

carrier protein. Any transport protein that alternates between two shapes to facilitate the transportation of an entity across the membrane.

Catabolism (or catabolic pathway). A metabolic pathway (set of chemical reactions) that releases energy by converting complex molecules to simpler molecules.

cell cycle. A series of events from the moment the cell forms from the division of its parent cell to the moment it divides itself into two daughter cells.

cell division. The process in which a parent cell divides into two or more daughter cells.

centriole. A barrel-shaped structure in a centrosome that helps the centrosome to organize microtubules into a spindle.

centrosome. A structure in the cytoplasm of animal cells that functions as a center to organize microtubules into a spindle. It usually includes two centrioles.

channel protein. A transport protein that provides a corridor (or channel) through which an entity (a specific molecule or ion) can cross the membrane.

chemiosmosis. The process of converting the energy stored in a proton gradient across the membrane to cellular energy such as energy in the form of ATP molecules.

clone. A copy of a gene or a genetically identical copy of a gene, cell, or organism.

cloning. The process of making a copy of a gene or a genetically identical copy of a gene, cell, or organism.

community. An ensemble of all populations of all species living close enough in the same area to have the potential of interacting with one another.

compound. A substance that is composed of two or more elements chemically mixed in a definite proportion.

covalent bond. A chemical bond between atoms made by sharing electrons.

cytokinesis. A process in which the cytoplasm of a parent cell under cell division is divided into two halves to form two daughter cells. It happens immediately after mitosis, meiosis I, or meiosis II.

cytoplasm. The content of a cell inside its plasma membrane or in between the plasma membrane and the nucleus in eukaryotic cells.

cytosol. The fluid component of cytoplasm.

decomposer (or detritivore). Any organism that obtains nutrients by decomposing wastes and remains of other organisms.

diffusion. The spontaneous movement of entities such as molecules down the concentration gradient.

DNA probe. A labeled single-stranded DNA fragment used in the lab to detect its complementary strand in a sample. When it runs into its complementary strand, it combines with it and sends back a signal.

DNA. Deoxyribose nucleic acid.

ecology. A subfield of biology in which scientists study the interaction between life and environment at different organizational (or complexity) levels such as organismal and population levels.

ecosystem. An ensemble of all the communities in a given region and the physical environment with which they interact.

electrochemical gradient. The gradient across the membrane formed by the difference in concentration and by the electric potential across the membrane.

electronegativity. The ability of an atom in a molecule to attract electrons toward itself.

electrogenic pump. A transport protein that generates electric potential across a membrane.

emergent properties. New properties at each organizational level of life which do not belong to any specific part.

endocytosis. The process by which a cell takes in large molecules and particulate matter by packaging it in vesicles made out of the plasma membrane of the cell.

eukaryotes. Organisms with eukaryotic cells.

eukaryotic cells. A type of cell that contains membrane-bound internal structures called organelles and a membrane-bound nucleus in which the DNA resides.

evolution. The emergence of lines of descent with modifications, that is, new species from common ancestors over generations of organisms in a population.

exocytosis. The process through which biological molecules are secreted out of the cell by packaging them in vesicles and then fusing the vesicles that carry them with the plasma membrane.

facilitated diffusion. Passive transport facilitated by membrane proteins; no energy expended.

gap junction. A junction formed by the membrane proteins of two adjacent cells which permits the rapid flow of ions and small enough molecules between the cytoplasms of the two cells.

genetic engineering. The application of scientific knowledge to manipulate genes for practical purposes such as developing components, systems, processes, and characteristics (traits).

genetic marker. An easily identifiable piece of genetic material, usually a DNA sequence that can be used to distinguish cells, individuals, populations, or species.

genetically modified organism (GMO) (or transgenic organism). An organism whose genome contains one or more genes from the genome of another organism of the same or different species.

genome. An entire set of genetic material in a single cell of an organism or in a virus. It includes both coding and noncoding sequences.

genomics. The study of the genomes of species.

heterotrophs. Organisms that obtain their food (organic molecules) by eating other organisms or the products derived from other organisms.

homeostasis. A steady state physiological condition of the body of a multicelled organism in which certain parameters are kept within a tolerable range.

integral protein. A membrane protein that either spans the whole membrane by penetrating through the hydrophobic core of the membrane or penetrates partway into the hydrophobic core from either side of the membrane.

ionic bond. A chemical bond between atoms that is formed when one atom loses one or more electrons and the other atom gains these electrons.

isomers. The compounds that have the same molecular formula but different structures.

metabolic pathway. A series of biochemical reactions that either builds a complex molecule from simpler ones or breaks down a complex molecule into smaller ones.

metabolism. Metabolism is the sum total of all the chemical reactions inside an organism, including reactions that make catabolic and anabolic pathways.

molecule. The smallest unit of a substance composed of two or more atoms bonded together with covalent bonds.

mutation. A lasting change in the nucleotide sequence of an organism's DNA or in the RNA of a virus.

operon. A self-regulated sequence of DNA containing a cluster of genes and an operator and a promoter that control gene expression.

Glossary

operator. A DNA segment (sequence of nucleotides) in and near the start of an operon to which a repressor binds in order to prevent the binding of a RNA polymerase to the promoter.

oxidation. Loss of one or more electrons by an atom in a chemical reaction.

oxidative phosphorylation. The process of synthesizing ATP by adding a phosphate group to an ADP in the third major stage of catabolism during which a large amount of cellular energy is generated in the form of ATP by using the energy derived from the redox reactions of an electron transport chain.

passive transport. The diffusion of entities along a biological membrane; no energy expended.

peripheral protein. A membrane protein that is loosely bound to the surface of the membrane and not embedded in it.

pH. The negative logarithm to the base ten of the hydrogen ion concentration in a solution.

phorylation. The process of adding a phosphate group to an organic molecule such as turning ADP into ATP.

photophosphorylaton. The process of synthesizing ATP by adding a phosphate group to an ADP during photosynthesis.

photosynthesis. The process that converts light energy to chemical energy by producing sugar and other organic molecules from water and carbon dioxide.

plasma membrane. The cellular membrane at the outer boundary of every cell which provides selective permeability and helps maintain an internal cellular environment that is different from the extracellular environment.

plasmid. A small circular DNA molecule with a few genes in it which exists in some bacteria in addition to chromosomes.

plasmid vector. A bacterial plasmid carrying genes from a source to a target (host).

polymerase chain reaction (PCR). A technique used to make copies of a DNA fragment in vitro with an exponentially amplified speed.

population. A group of organisms of the same species living in the same area at the same time.

pre-mRNA (or primary transcript or primary mRNA). The initial RNA transcript (before splicing) in a eukaryotic cell synthesized by using the genetic instructions in a DNA template strand.

prokaryotes. Organisms with prokaryotic cells.

prokaryotic cells. A type of cell that lacks membrane-bound internal structures called organelles and a membrane-bound nucleus.

promoter. A specific sequence of nucleotides in a DNA to which RNA polymerase binds in order to initiate transcription at a nearby gene.

proton pump. A protein that actively transports the hydrogen ions through the membrane from the cytoplasm to the extracellular fluid against the electrochemical gradient.

redox (oxidation-reduction) reaction. Any chemical reaction in which one or more electrons completely or partially transfer from one atom to another.

recombinant DNA. A DNA molecule that contains segments of genetic material from different sources.

recombinant DNA technology. The in vitro molecular technique to isolate and manipulate fragments of DNA.

reduction. The gain of one or more electrons by an atom in a chemical reaction.

repressor. A protein that binds to an appropriate place on a DNA to inhibit gene transcription, for example, by preventing RNA polymerase from binding to the promoter.

restriction enzyme. An enzyme that recognizes and cuts a DNA molecule at specific nucleotide sequences.

restriction site. A specific nucleotide sequence on a DNA molecule or strand that can be recognized and cut by a restriction enzyme.

RNA interference (RNA-i). A technique used to silence the expression of one or more selected genes.

RNA splicing. The process of synthesizing mRNA from a pre-mRNA by removing portions of it called introns and joining together the rest of the portions called exons.

species. One or more groups of organisms that have the potential to interbreed and produce viable and fertile offspring and do not have the same potential to interbreed and reproduce with members of other groups.

sexual reproduction. Reproduction in which the offspring inherits half of its genes from each of two parents.

Spindle (or bipolar spindle). A dynamic array of microtubules with two poles directing the motion of chromosomes during mitosis and meiosis.

tight junction. A junction formed by the membrane proteins of two adjacent cells which prevents the flow of material between the cytoplasms of the two cells.

transcription. The process used to synthesize a RNA from a DNA template strand.

transcription factors. Regulatory proteins that influence transcription in various ways such as activating it or repressing it.

transcription initiation complex. A collection of RNA polymerase and transcription factors bound to a promoter.

transgenic organism (or genetically modified organism or GMO). An organism whose genome contains one or more genes from the genome of another organism of the same or different species.

translation. The process of synthesizing a polypeptide chain by using the genetic information in an mRNA.

vector. An organism that spreads pathogens from one organism to another.

Credits and Acknowledgments

Unless otherwise acknowledged, all pictures and illustrations in this book are the property of Infonential, Inc. We have made our best effort to trace and acknowledge the ownership of the following items. In the event of any question or issue arising from the use of any of these items, we will be pleased to make the necessary corrections in future printings.

Chapter 1. Figure 2.1 GNU Free Documentation License. Figures 2.3 and 2.4 are in the public domain.

Chapter 3. Figures 3.4, 3.5, 3.7, 3.8, and 3.9 are in the public domain.

Chapter 4. Figures 4.1 and 4.2 are courtesy of Mariana Ruiz Villarrea. Figure 4.3 is courtesy of Messer Woland. Figure 4.4 is in the public domain.

Chapter 5. Figures 5.1 and 5.3 are courtesy of Mariana Ruiz Villarrea. Figure 5.4 is in the public domain.

Chapter 6. Illustration of ATP molecule and Figures 6.1, 6.2, 6.10, 6.12, and 6.14 are based on figures in the public domain.

Chapter 7. Figures 7.1 and 7.2 are courtesy of Mariana Ruiz Villarrea. Figure 7.3 is in the public domain. Figure 7.4 is courtesy of the National Center for Biotechnology Information.

Chapter 8. Figure 8.1 is courtesy of the American Philosophical Society. Figures 8.3 and 8.4 are in the public domain.

Chapter 9. Figure 9.3 is courtesy of Madeleine Price Ball. Figures 9.4, 9.5, 9.6, 9.10, and 9.15 are in the public domain.

Molecular and Cellular Biology

Figure 9.8 is courtesy of Daniel Horspool. Figure 9.12 is courtesy of Mariana Ruiz Villarrea. Figure 9.14 GNU Free Documentation License.

Chapter 10. Figures 10.2 and 10.8 are based on images in the public domain. Figure 10.3 is courtesy of Magnus Manske.